SOLUTIONS MANUAL FOR
MATHEMATICAL STRUCTURES
FOR COMPUTER SCIENCE

SOLUTIONS MANUAL FOR
MATHEMATICAL STRUCTURES FOR COMPUTER SCIENCE

SECOND EDITION

JUDITH L. GERSTING

W. H. FREEMAN AND COMPANY
NEW YORK

Cover design adapted from a graphic by Benoit B. Mandelbrot and reproduced by his kind permission. From *The Fractal Geometry of Nature* by Benoit B. Mandelbrot. W. H. Freeman and Company. Copyright © 1982.

Copyright © 1987 by W. H. Freeman and Company

No part of this book may be reproduced by any mechanical, photographic, or electronic process, or in the form of a photographic recording, nor may it be stored in a retrieval system, transmitted, or otherwise copied for public or private use, without written permission of the publisher.

Printed in the United States of America

2 3 4 5 6 7 8 9 0   SL   5 4 3 2 1 0 8 9 8 7

ISBN 0-7167-1803-0

CONTENTS

CHAPTER 1 1
CHAPTER 2 41
CHAPTER 3 55
CHAPTER 4 68
CHAPTER 5 88
CHAPTER 6 97
CHAPTER 7 113
CHAPTER 8 135
CHAPTER 9 145
CHAPTER 10 159
CHAPTER 11 182
CHAPTER 12 192

Answers that are starred also appear at the back of the textbook.

CHAPTER 1

*1. (a), (c), (e), (f)

2. a) T b) F c) F d) F e) T f) F g) T h) T

*3. a) antecedent: sufficient water
       consequent: healthy plant growth
    b) antecedent: further technological advances
       consequent: increased availability of microcomputers
    c) antecedent: errors will be introduced
       consequent: there is a modification of the program
    d) antecedent: fuel savings
       consequent: good insulation or all windows are storm windows

4. a) 1 and 3   b) 2   c) 4

5. a) $A \wedge B$
   b) $A \wedge (B \vee C)$
   c) $B \rightarrow (A \wedge C)$
   d) $A \rightarrow (B' \vee C')$
   e) $A \wedge (C' \rightarrow (B' \vee C))$

6. a) Violets are blue or sugar is sour.
   b) Violets are not blue or, if roses are red, then sugar is sweet.
   c) Sugar is sweet and roses are not red, if and only if violets are blue.
   d) Sugar is sweet, and roses are not red if and only if violets are blue.
   e) If it is false that both violets are blue and sugar is sour, then roses are red.
   f) Roses are red, or violets are blue and sugar is sour.
   g) Roses are red or violets are blue, and sugar is sour.

*7. a) A: prices go up; B: housing will be plentiful;
       C: housing will be expensive
       $[A \rightarrow B \wedge C] \wedge (C' \rightarrow B)$
    b) A: going to bed; B: going swimming; C: changing clothes
       $[(A \vee B) \rightarrow C] \wedge (C \rightarrow B)'$
    c) A: it will rain; B: it will snow
       $(A \vee B) \wedge (A \wedge B)'$

d) A: Janet wins; B: Janet loses; C: Janet will be tired
   $(A \lor B) \rightarrow C$

e) A: Janet wins; B: Janet loses; C: Janet will be tired
   $A \lor (B \rightarrow C)$

8. *a)

A	B	A→B	A'	A' ∨ B	(A→B) ↔ A' ∨ B
T	T	T	F	T	T
T	F	F	F	F	T
F	T	T	T	T	T
F	F	T	T	T	T

Tautology

*b)

A	B	C	A ∧ B	(A ∧ B) ∨ C	B ∨ C	A ∧ (B ∨ C)
T	T	T	T	T	T	T
T	T	F	T	T	T	T
T	F	T	F	T	T	T
T	F	F	F	F	F	F
F	T	T	F	T	T	F
F	T	F	F	F	T	F
F	F	T	F	T	T	F
F	F	F	F	F	F	F

$(A \land B) \lor C \rightarrow A \land (B \lor C)$

T
T
T
T
F
T
F
T

c)

A	B	A'	B'	A' ∨ B'	(A' ∨ B')'	A ∧ (A' ∨ B')'
T	T	F	F	F	T	T
T	F	F	T	T	F	F
F	T	T	F	T	F	F
F	F	T	T	T	F	F

d)

A	B	A'	A ∧ B	A ∧ B → A'
T	T	F	T	F
T	F	F	F	T
F	T	T	F	T
F	F	T	F	T

e)

A	B	C	A → B	A ∨ C	B ∨ C	(A ∨ C) → (B ∨ C)
T	T	T	T	T	T	T
T	T	F	T	T	T	T
T	F	T	F	T	T	T
T	F	F	F	T	F	F
F	T	T	T	T	T	T
F	T	F	T	F	T	T
F	F	T	T	T	T	T
F	F	F	T	F	F	T

(A → B) → [(A ∨ C) → (B ∨ C)]
T
T
T
T
T
T
T
T

Tautology

f)

A	B	B → A	A → (B → A)
T	T	T	T
T	F	T	T
F	T	F	T
F	F	T	T

Tautology

g)

A	B	A ∧ B	A'	B'	B' ∨ A'	A ∧ B ↔ B' ∨ A'
T	T	T	F	F	F	F
T	F	F	F	T	T	F
F	T	F	T	F	T	F
F	F	F	T	T	T	F

Contradiction

h)

A	B	B'	A ∨ B'	A ∧ B	(A ∧ B)'	(A ∨ B') ∧ (A ∧ B)'
T	T	F	T	T	F	F
T	F	T	T	F	T	T
F	T	F	F	F	T	F
F	F	T	T	F	T	T

i)

A	B	C	A ∨ B	C'	(A ∨ B) ∧ C'	A'	A' ∨ C
T	T	T	T	F	F	F	T
T	T	F	T	T	T	F	F
T	F	T	T	F	F	F	T
T	F	F	T	T	T	F	F
F	T	T	T	F	F	T	T
F	T	F	T	T	T	T	T
F	F	T	F	F	F	T	T
F	F	F	F	T	F	T	T

[(A ∨ B) ∧ C'] → A' ∨ C

T
F
T
F
T
T
T
T

*9. $2^{2^4} = 2^{16}$

10. 1.0, 2.4, 7.2, 5.3

11. For example: (A OR B) AND NOT (A AND B) AND NOT C

12. 1b)

A	B	A ∧ B	B ∧ A	A ∧ B ↔ B ∧ A
T	T	T	T	T
T	F	F	F	T
F	T	F	F	T
F	F	F	F	T

2a)

A	B	C	A ∨ B	(A ∨ B) ∨ C	B ∨ C	A ∨ (B ∨ C)
T	T	T	T	T	T	T
T	T	F	T	T	T	T
T	F	T	T	T	T	T
T	F	F	T	T	F	T
F	T	T	T	T	T	T
F	T	F	T	T	T	T
F	F	T	F	T	T	T
F	F	F	F	F	F	F

(A ∨ B) ∨ C ↔ A ∨ (B ∨ C)
T
T
T
T
T
T
T
T

2b)

A	B	C	A ∧ B	(A ∧ B) ∧ C	B ∧ C	A ∧ (B ∧ C)
T	T	T	T	T	T	T
T	T	F	T	F	F	F
T	F	T	F	F	F	F
T	F	F	F	F	F	F
F	T	T	F	F	T	F
F	T	F	F	F	F	F
F	F	T	F	F	F	F
F	F	F	F	F	F	F

(A ∧ B) ∧ C ↔ A ∧ (B ∧ C)
T
T
T
T
T
T
T
T

3a)

A	B	C	B ∧ C	A ∨ (B ∧ C)	A ∨ B	A ∨ C
T	T	T	T	T	T	T
T	T	F	F	T	T	T
T	F	T	F	T	T	T
T	F	F	F	T	T	T
F	T	T	T	T	T	T
F	T	F	F	F	T	F
F	F	T	F	F	F	T
F	F	F	F	F	F	F

(A ∨ B) ∧ (A ∨ C)	A ∨ (B ∧ C) ↔ (A ∨ B) ∧ (A ∨ C)
T	T
T	T
T	T
T	T
T	T
F	T
F	T
F	T

3b)

A	B	C	B ∨ C	A ∧ (B ∨ C)	A ∧ B	A ∧ C
T	T	T	T	T	T	T
T	T	F	T	T	T	F
T	F	T	T	T	F	T
T	F	F	F	F	F	F
F	T	T	T	F	F	F
F	T	F	T	F	F	F
F	F	T	T	F	F	F
F	F	F	F	F	F	F

(A ∧ B) ∨ (A ∧ C)	A ∧ (B ∨ C) ↔ (A ∧ B) ∨ (A ∧ C)
T	T
T	T
T	T
F	T
F	T
F	T
F	T
F	T

4a)
A	0	A∨0	A∨0 ⟷ A
T	F	T	T
F	F	F	T

5b)
A	A'	A∧A'	0	A∧A' ⟷ 0
T	F	F	F	T
F	T	F	F	T

13. *a)
A	A'	A ∨ A'
T	F	T
F	T	T

b)
A	A'	(A')'	(A')' ⟷ A
T	F	T	T
F	T	F	T

*c)
A	B	A∧B	A∧B → B
T	T	T	T
T	F	F	T
F	T	F	T
F	F	F	T

d)
A	B	A∨B	A → A∨B
T	T	T	T
T	F	T	T
F	T	T	T
F	F	F	T

e)
A	B	A∨B	(A∨B)'	A'	B'	A'∧B'	(A∨B)' ⟷ A'∧B'
T	T	T	F	F	F	F	T
T	F	T	F	F	T	F	T
F	T	T	F	T	F	F	T
F	F	F	T	T	T	T	T

f)
A	B	A∧B	(A∧B)'	A'	B'	A'∨B'	(A∧B)' ⟷ A'∨B'
T	T	T	F	F	F	F	T
T	F	F	T	F	T	T	T
F	T	F	T	T	F	T	T
F	F	F	T	T	T	T	T

14.*a) Assign

    B' ∧ (A → B)   true
    A'   false

From the second assignment, A is true. From the first assignment, B' is true (so B is false), and A → B is true. If A → B is true and A is true, then B is true. B is thus both true and false, and (B' ∧ (A → B)) → A' is a tautology.

b) Assign

$(A \rightarrow B) \wedge A$ true
B false

From the first assignment, A is true and A → B is true. If A → B is true and A is true, then B is true. B is thus both true and false, and $[(A \rightarrow B) \wedge A] \rightarrow B$ is a tautology.

c) Assign

$(A \vee B) \wedge A'$ true
B false

From the first assignment, A' is true (and A is false), and A ∨ B is true. If A ∨ B is true and A is false, then B is true. B is thus both true and false, and $(A \vee B) \wedge A' \rightarrow B$ is a tautology.

d) Assign

$(A \wedge B) \wedge B'$ true
A false

From the first assignment, A ∧ B is true. If A ∧ B is true, then A is true. A is thus both true and false, and $(A \wedge B) \wedge B' \rightarrow A$ is a tautology.

15. a) P is A ∨ A'   Q is B ∨ B'
    b) P is A ∧ A'
    c) P is A ∨ A'   Q is A ∧ A'

16. a)

A	B	A ⊕ B
T	T	F
T	F	T
F	T	T
F	F	F

b)

A	B	A ⊕ B	A ↔ B	(A ↔ B)'	A ⊕ B ↔ (A ↔ B)'
T	T	F	T	F	T
T	F	T	F	T	T
F	T	T	F	T	T
F	F	F	T	F	T

17. a)

A	B	A ∨ B	A'	B'	A' ∧ B'	(A' ∧ B')'	A ∨ B ↔ (A' ∧ B')'
T	T	T	F	F	F	T	T
T	F	T	F	T	F	T	T
F	T	T	T	F	F	T	T
F	F	F	T	T	T	F	T

b)

A	B	B'	A ∧ B'	(A ∧ B')'	A → B	A → B ↔ (A ∧ B')'
T	T	F	F	T	T	T
T	F	T	T	F	F	T
F	T	F	F	T	T	T
F	F	T	F	T	T	T

18. a) A ∧ B is equivalent to (A' ∨ B')'

A	B	A ∧ B	A'	B'	A' ∨ B'	(A' ∨ B')'	A ∧ B ↔ (A' ∨ B')'
T	T	T	F	F	F	T	T
T	F	F	F	T	T	F	T
F	T	F	T	F	T	F	T
F	F	F	T	T	T	F	T

A → B is equivalent to A' ∨ B

A	B	A → B	A'	A' ∨ B	A → B ↔ A' ∨ B
T	T	T	F	T	T
T	F	F	F	F	T
F	T	T	T	T	T
F	F	T	T	T	T

b) A ∧ B is equivalent to (A → B')'

A	B	A ∧ B	B'	A → B'	(A → B')'	A ∧ B ↔ (A → B')'
T	T	T	F	F	T	T
T	F	F	T	T	F	T
F	T	F	F	T	F	T
F	F	F	T	T	F	T

A ∨ B is equivalent to A' → B

A	B	A ∨ B	A'	A' → B	A ∨ B ↔ A' → B
T	T	T	F	T	T
T	F	T	F	T	T
F	T	T	T	T	T
F	F	F	T	F	T

19. (A ∧ B)' has the value F when A and B have the values T. However, any statement using only → and ∨ will have the value T when A and B are both T.

20. A ∧ B is equivalent to (A|B)|(A|B)

A	B	A ∧ B	A|B	(A|B)|(A|B)	A ∧ B ↔ (A|B)|(A|B)
T	T	T	F	T	T
T	F	F	T	F	T
F	T	F	T	F	T
F	F	F	T	F	T

A' is equivalent to A|A

A	A'	A|A	A' ↔ A|A
T	F	F	T
F	T	T	T

21. A ∧ B is equivalent to (A ↓ A) ↓ (B ↓ B)

A	B	A ∧ B	A ↓ A	B ↓ B	(A ↓ A) ↓ (B ↓ B)
T	T	T	F	F	T
T	F	F	F	T	F
F	T	F	T	F	F
F	F	F	T	T	F

A ∧ B ↔ (A ↓ A) ↓ (B ↓ B)

T
T
T
T

A' is equivalent to A ↓ A

A	A'	A ↓ A	A' ↔ A ↓ A
T	F	F	T
F	T	T	T

22. Percival's statement is of the form A → B, where A stands for "I am a truth-teller" and B stands for "Llewellyn is a truth-teller." There are two possibilities: A is T or A is F. If A is T, then A → B is T because Percival said it. With A T and A → B T, B is T. If A is F, then the implication A → B is T but Percival, as a liar, would not have said a true statement. Therefore this possibility can't happen, and Percival and Llewellyn are both truth-tellers.

23. * a) true (pick y = 0)
    * b) true (pick y = 0)
    * c) true (pick y = -x)
    * d) false (no one y works for all x's)

      e) false (may have x = y)
      f) true (pick y = -x)
      g) true (pick x = 2, y = 4)
      h) false (may have x = 0)

24. a) F b) T c) T d) F e) T

25. * a) true: domain is the set of integers, $A(x)$ is "x is even", $B(x)$ is "x is odd"
      false: domain is the positive integers, $A(x)$ is "x > 0", $B(x)$ is "x ≥ 1"
   b) true: domain is the set of lines in the plane, $P(x,y)$ is "x is parallel to y"
      false: domain is the set of integers, $P(x,y)$ is "x < y"
   c) true: domain is the set of integers, $P(x)$ is "x is even", $Q(x,y)$ is "y|x" (y divides x)
      false: domain is the set of all people, $P(x)$ is "x is male", $Q(x,y)$ is "y is a brother of x"
   d) true: domain is the set of nonnegative integers, $A(x)$ is "x is even", $B(x,y)$ is "x ≤ y"
      false: domain is the set of positive integers, $A(x)$ is "x is even", $B(x,y)$ is "x ≤ y"
   e) true: domain is the set of integers, $A(x)$ is "x > 0", $B(x)$ is "x ≥ 0"
      false: domain is the set of integers, $A(x)$ is "x > 0", $B(x)$ is "x is even"

26. a) scope of $(\forall x)$ is $P(x) \rightarrow Q(y)$; y is a free variable
   b) scope of $(\exists x)$ is $A(x) \land (\forall y)B(y)$; scope of $(\forall y)$ is $B(y)$; no free variables
   c) scope of $(\exists x)$ is $(\forall y)P(x,y) \land Q(x,y)$; scope of $(\forall y)$ is $P(x,y)$; y is a free variable
   d) scope of $(\exists x)$ is $(\exists y)(A(x,y) \land B(y,z) \rightarrow A(a,z))$; scope of $(\exists y)$ is $A(x,y) \land B(y,z) \rightarrow A(a,z)$; z is a free variable

27. * a) $(\forall x)(D(x) \rightarrow S(x))$
  * b) $(\exists x)(D(x) \land (R(x))')$
  * c) $(\forall x)(D(x) \land S(x) \rightarrow (R(x))')$
    d) $(\exists x)(D(x) \land S(x) \land R(x))$
    e) $(\forall x)(D(x) \rightarrow (S(x) \land R(x))')$
    f) $(\forall x)(D(x) \land S(x) \rightarrow D(x) \land R(x))$

g) $(\forall x)(D(x) \rightarrow (S(x))')$
h) $S(M) \rightarrow (\forall x)(D(x) \rightarrow S(x))$
i) $R(M) \land R(T)$
j) $(\exists x)(D(x) \land R(x)) \rightarrow (\forall x)(D(x) \rightarrow S(x))$

28. * a) $(\forall x)(C(x) \land F(x))'$
    b) $(\exists x)[P(x) \land (\forall y)(S(x,y) \rightarrow F(y))]$
 * c) $(\forall x)(\forall y)[(P(y) \land S(x,y)) \rightarrow C(x)]$
    d) $(\forall x)[F(x) \rightarrow (\exists y)(C(y) \land S(x,y))]$
    e) $(\exists x)[P(x) \land (\forall y)(C(y) \rightarrow (S(x,y))')]$
    f) $(\exists x)(\exists y)(C(x) \land F(y) \land S(x,y)) \rightarrow$
       $(\forall x)(\forall y)(C(x) \land F(y) \rightarrow S(x,y))$

29. * a) John is handsome and Kathy loves John
  * b) all men are handsome
    c) all women love only handsome men
    d) a handsome man loves Kathy
    e) some pretty woman loves only handsome men
    f) John loves all pretty women

30. a) both sides are true exactly when A(x,y) holds for all x,y pairs
    b) both sides are true exactly when some x,y pair satisfies the property A(x,y).
    c) if there is a single x that is in relation P to all y, then for every y an x exists (this same x) that is in relation P to y
    d) if a has property A, then something in the domain has property A.
    e) if any member of the domain that has property A also has property B, then if all members of the domain have property A, all have property B.

31. * a) domain is the set of integers, A(x) is "x is even", B(x) is "x is odd"
    b) domain is the set of integers, P(x,y) is "x + y = 0"; for every x there is a y (y = -x) such that x + y = 0 but there is no single integer x that gives 0 when added to every integer y.

c) domain is the set of positive integers, P(x) is "x > 4", Q(x) is "x > 2". Then every positive integer greater than 4 is greater than 2, so $(\forall x)(P(x) \rightarrow Q(x))$ is true. There exists a positive integer greater than 4, but not all positive integers are greater than 2, so $(\exists x)P(x) \rightarrow (\forall x)Q(x)$ is false.

d) domain is the set of integers, A(x) is "x is even". Then $(\forall x)(A(x))'$ is false - it is not the case that every integer is odd (not even) - but $((\forall x)A(x))'$ is true since it is false that every integer is even.

32. a) valid: there is an x in the domain with property A says it is false that everything in the domain fails to have property A.

b) not valid: domain is the set of integers, P(x) is "x is even", Q(x) is "x is prime". Because there are prime integers, $(\exists x)Q(x)$ and therefore $(\forall x)P(x) \vee (\exists x)Q(x)$ is true. But it is false that every integer is even or prime, so the implication is false.

c) valid: A true for all objects in the domain means it is false that there is some object in the domain for which A is not true.

d) valid: suppose that for every member of the domain, either P(x) or Q(x) is true. If there is some member of the domain for which Q is true, then $(\exists y)Q(y)$ is true. Otherwise all members of the domain have property P and $(\forall x)P(x)$ is true. In either case, $(\forall x)P(x) \vee (\exists y)Q(y)$ is true.

33. a) 2   b) 3   c) 3   d) 1

34. a) hamburger, cheeseburger, fries, fish

b) milk

c) Frosty, rootbeer, Pepsi, milk, Coke, lemonade, tea, milk (milk gets listed twice)

d) cheeseburger, fish

Section 1.2

*1. 1. hypothesis
    2. hypothesis
    3. Axiom 2
    4. 2,3,modus ponens
    5. 1,4,modus ponens

 2. 1. 1.25
    2. Axiom 2
    3. 1,2,modus ponens

*3. 1. (P')'   hypothesis
    2. (P')' → [P' → ((P')')']   1.27
    3. P' → ((P')')'   1,2,modus ponens
    4. [P' → ((P')')'] → [(P')' → P]   Axiom 3
    5. (P')' → P   3,4,modus ponens
    6. P   1,5,modus ponens

 4. 1. ((P')')' → P'   Exercise 3
    2. [((P')')' → P'] → (P → (P')')   Axiom 3
    3. P → (P')'   1,2,modus ponens

 5. Eliminating the connective ∨, we want to prove P → (P' → Q).
    1. P   hypothesis
    2. P'   hypothesis
    3. P' → (P → Q)   1.27
    4. P → Q   2,3,modus ponens
    5. Q   1,4,modus ponens

 6. 1. P → Q   hypothesis
    2. (P')' → P   Exercise 3
    3. ((P')' → P) ∧ (P → Q)   A ∧ B can be deduced from A,B
    4. (P')' → Q   3,1.29,modus ponens
    5. Q → (Q')'   Exercise 4
    6. ((P')' → Q) ∧ (Q → (Q')')   A ∧ B can be deduced from A,B
    7. (P')' → (Q')'   6,1.29,modus ponens
    8. [(P')' → (Q')'] → (Q' → P')   Axiom 3
    9. Q' → P'   7,8,modus ponens

 7. Eliminating the connective ∨, we want to prove
    (P' → Q) → (Q' → P)

1. P' → Q   hypothesis
2. (P' → Q) → (Q' → (P')')   Exercise 6
3. Q' → (P')'   1,2,modus ponens
4. Q'   hypothesis
5. (P')'   1,4,modus ponens
6. (P')' → P   Exercise 3
7. P   5,6,modus ponens

8. 1. P → (Q → R)   hypothesis
   2. P → (Q → R) → ((P → Q) → (P → R))   Axiom 2
   3. (P → Q) → (P → R)   1,2,modus ponens
   4. ((P → Q) → (P → R)) →
      [Q → ((P → Q) → (P → R))]   Axiom 1
   5. Q → ((P → Q) → (P → R))   3,4,modus ponens
   6. [Q → ((P → Q) → (P → R))] →
      [(Q → (P → Q)) → (Q → (P → R))]   Axiom 2
   7. (Q → (P → Q)) → (Q → (P → R)) 5,6,modus ponens
   8. Q → (P → Q)   Axiom 1
   9. Q → (P → R)   7,8,modus ponens

*9. Eliminating the connective ∨, we want to prove
    P' ∧ (P' → Q) → Q
   1. P'   hypothesis
   2. P' → Q   hypothesis
   3. Q   1,2,modus ponens

10. 1. P'   hypothesis
    2. Q → P   hypothesis
    3. (Q → P) → (P' → Q')   Exercise 6
    4. P' → Q'   2,3,modus ponens
    5. Q'   1,4,modus ponens

11. The argument is: (E → Q) ∧ (E ∨ B) ∧ Q' → B
    A proof sequence is:
    1. E → Q   hypothesis
    2. E ∨ B   hypothesis
    3. Q'   hypothesis
    4. E' → B   substitution from equivalence
                                    (E ∨ B) ↔ (E' → B) into 2
    5. (E → Q) → (Q' → E')   Exercise 6
    6. Q' → E'   1,5,modus ponens

15

    7. E'   3,6,modus ponens
    8. B   4,7,modus ponens

12. The argument is: $(C \land W') \land ((R \lor S') \to W) \to C \land S$
    A proof sequence is:
    1. $C \land W'$ hypothesis
    2. $(R \lor S') \to W$ hypothesis
    3. W' 1,tautology $A \land B \to B$, modus ponens
    4. $((R \lor S') \to W) \to (W' \to (R \lor S')')$ Exercise 6
    5. $W' \to (R \lor S')'$ 2,4,modus ponens
    6. $(R \lor S')'$ 3,5,modus ponens
    7. $(R \lor S')' \to R' \land S$ tautology
    8. $R' \land S$ 6,7,modus ponens
    9. C 1,tautology $A \land B \to A$, modus ponens
    10. S 8,tautology $A \land B \to B$, modus ponens
    11. $C \land S$ $A \land B$ can be deduced from A,B

*13. The argument is: $(R \land (F' \lor N)) \land N' \land (A' \to F) \to A \land R$
    A proof sequence is:
    1. $R \land (F' \lor N)$ hypothesis
    2. N' hypothesis
    3. $A' \to F$ hypothesis
    4. R 1,tautology $A \land B \to A$, modus ponens
    5. $F' \lor N$ 1,tautology $A \land B \to B$, modus ponens
    6. $(F' \lor N) \land N'$ $A \land B$ can be deduced from A,B
    7. $(F' \lor N) \land N' \to F'$ tautology
    8. F' 6,7,modus ponens
    9. $(A' \to F) \to (F' \to A)$ tautology
    10. $F' \to A$ 3,9,modus ponens
    11. A 8,10,modus ponens
    12. $A \land R$ $A \land B$ can be deduced from A,B

14. The argument is: $(R \to U)' \land (P \lor B')' \to U' \land B$
    A proof sequence is:
    1. $(R \to U)'$ hypothesis
    2. $(R \to U)' \to R \land U'$ tautology
    3. $R \land U'$ 1,2,modus ponens
    4. $(P \lor B')'$ hypothesis
    5. $(P \lor B')' \to P' \land B$ tautology
    6. $P' \land B$ 4,5,modus ponens

     7. U'   3, tautology $A \wedge B \rightarrow B$, modus ponens
     8. B   6, tautology $A \wedge B \rightarrow B$, modus ponens
     9. U' $\wedge$ B $A \wedge B$ can be deduced from A, B

*15. 1. $(\forall x)P(x)$ hypothesis
    2. $P(x)$ 1, Axiom 5, modus ponens
    3. $P(x) \rightarrow P(x) \vee Q(x)$ tautology
    4. $P(x) \vee Q(x)$ 2, 3, modus ponens
    5. $(\forall x)(P(x) \vee Q(x))$ 4, generalization (note that
                                  $P(x) \vee Q(x)$ was deduced from
                                  $(\forall x)P(x)$, in which x is not free)

16. 1. $(\forall x)P(x)$ hypothesis
    2. $(\exists x)Q(x)$ hypothesis
    3. $Q(a)$ 2, Axiom 6, modus ponens
    4. $P(a)$ 1, Axiom 5, modus ponens
    5. $Q(a) \wedge P(a)$ $A \wedge B$ can be deduced from A, B
    6. $(\exists x)(Q(x) \wedge P(x))$ 5, Axiom 7, modus ponens

17. 1. $(\exists x)(\exists y)P(x,y)$ hypothesis
    2. $(\exists y)P(a,y)$ 1, Axiom 6, modus ponens
    3. $P(a,b)$ 2, Axiom 6, modus ponens
    4. $(\exists x)P(x,b)$ 3, Axiom 7, modus ponens
    5. $(\exists y)(\exists x)P(x,y)$ 4, Axiom 7, modus ponens

18. 1. $(\forall x)(\forall y)Q(x,y)$ hypothesis
    2. $(\forall y)Q(x,y)$ 1, Axiom 5, modus ponens
    3. $Q(x,y)$ 2, Axiom 5, modus ponens
    4. $(\forall x)Q(x,y)$ 3, generalization (x not free in
                                  $(\forall x)(\forall y)Q(x,y)$)
    5. $(\forall y)(\forall x)Q(x,y)$ 4, generalization (y not free in
                                  $(\forall x)(\forall y)Q(x,y)$)

*19. 1. $(\exists x)(A(x) \wedge B(x))$ hypothesis
    2. $A(a) \wedge B(a)$ 1, Axiom 6, modus ponens
    3. $A(a)$ 2, tautology $A \wedge B \rightarrow A$, modus ponens
    4. $B(a)$ 2, tautology $A \wedge B \rightarrow B$, modus ponens
    5. $(\exists x)A(x)$ 3, Axiom 7, modus ponens
    6. $(\exists x)B(x)$ 4, Axiom 7, modus ponens
    7. $(\exists x)A(x) \wedge (\exists x)B(x)$ $A \wedge B$ can be deduced from A, B

20. Eliminating the second disjunction, we want to prove
    $(\exists x)(R(x) \lor S(x)) \rightarrow [((\exists x)R(x))' \rightarrow (\exists x)S(x)]$
    1. $(\exists x)(R(x) \lor S(x))$ hypothesis
    2. $R(a) \lor S(a)$ 1, Axiom 6, modus ponens
    3. $((\exists x)R(x))'$ hypothesis
    4. $((\exists x)R(x))' \leftrightarrow (\forall x)(R(x))'$ Axiom 8
    5. $(\forall x)(R(x))'$ substitution from 4 into 3
    6. $(R(a))'$ 5, Axiom 5, modus ponens
    7. $S(a)$ 2,6, tautology $(A \lor B) \land A' \rightarrow B$, modus ponens
    8. $(\exists x)S(x)$ 7, Axiom 7, modus ponens

21. $(\exists x)(\forall y)Q(x,y) \rightarrow (\forall y)(\exists x)Q(x,y)$
    1. $(\exists x)(\forall y)Q(x,y)$ hypothesis
    2. $(\forall y)Q(a,y)$ 1, Axiom 6, modus ponens
    3. $Q(a,y)$ 2, Axiom 5, modus ponens
    4. $(\exists x)Q(x,y)$ 3, Axiom 7, modus ponens
    5. $(\forall y)(\exists x)Q(x,y)$ 4, generalization (conditions a and b hold)

22. 1. $(\forall x)(A(x) \rightarrow B(x))$ hypothesis
    2. $(\exists x)A(x)$ hypothesis
    3. $A(a)$ 2, Axiom 6, modus ponens
    4. $A(a) \rightarrow B(a)$ 1, Axiom 5, modus ponens
    5. $B(a)$ 3,4, modus ponens
    6. $(\exists x)B(x)$ 5, Axiom 7, modus ponens

23. The use of generalization at step 4 violates condition (b) because at step 2, y is free in $(\exists x)Q(x,y)$, and then Axiom 6 was applied at step 3.

24. The argument is:
    $(\exists x)(A(x) \land (N(x))') \land (\forall x)(G(x) \rightarrow N(x)) \land (\forall x)(G(x) \lor C(x)) \rightarrow (\exists x)(A(x) \land C(x))$
    A proof sequence is:
    1. $(\exists x)(A(x) \land (N(x))')$ hypothesis
    2. $A(a) \land (N(a))'$ 1, Axiom 6, modus ponens
    3. $(\forall x)(G(x) \rightarrow N(x))$ hypothesis
    4. $G(a) \rightarrow N(a)$ 3, Axiom 5, modus ponens
    5. $(G(a) \rightarrow N(a)) \rightarrow [(N(a))' \rightarrow (G(a))']$ tautology (or Exercise 6)
    6. $(N(a))' \rightarrow (G(a))'$ 4,5, modus ponens
    7. $(N(a))'$ 2, tautology $A \land B \rightarrow B$, modus ponens

8. (G(a))'   6,7,modus ponens
9. (∀x)(G(x) ∨ C(x))   hypothesis
10. G(a) ∨ C(a)   9,Axiom 5,modus ponens
11. (G(a))' ∧ (G(a) ∨ C(a)) → C(a)   tautology
12. C(a)   8,10,11,modus ponens
13. A(a)   2,tautology A∧B→A,modus ponens
14. A(a) ∧ C(a)   A∧B can be deduced from A,B
15. (∃x)(A(x) ∧ C(x))   14,Axiom 7, modus ponens

*25. The argument is:

(∀x)(M(x) → I(x) ∨ G(x)) ∧ (∀x)(G(x) ∧ L(x) → F(x)) ∧ (I(j))' ∧ L(j) → (M(j) → F(j))

A proof sequence is:

1. (∀x)(M(x) → I(x) ∨ G(x))   hypothesis
2. (∀x)(G(x) ∧ L(x) → F(x))   hypothesis
3. M(j) → I(j) ∨ G(j)   1,Axiom 5, modus ponens
4. G(j) ∧ L(j) → F(j)   2,Axiom 5, modus ponens
5. M(j)   hypothesis
6. I(j) ∨ G(j)   3,5, modus ponens
7. (I(j))'   hypothesis
8. (I(j))' ∧ (I(j) ∨ G(j)) → G(j)   tautology
9. G(j)   6,7,8, modus ponens
10. L(j)   hypothesis
11. G(j) ∧ L(j)   A ∧ B can be deduced from A,B
12. F(j)   4,11, modus ponens

26. The argument is:

(∃x)(M(x) ∧ (∀y)R(x,y)) ∧ (∀x)(∀y)(R(x,y) → T(x,y)) → (∃x)(M(x) ∧ (∀y)T(x,y))

A proof sequence is:

1. (∃x)(M(x) ∧ (∀y)R(x,y))   hypothesis
2. M(a) ∧ (∀y)R(a,y)   1, Axiom 6, modus ponens
3. M(a)   2,tautology A∧B→A,modus ponens
4. (∀x)(∀y)(R(x,y) → T(x,y))   hypothesis
5. (∀y)(R(a,y) → T(a,y))   4, Axiom 5, modus ponens
6. (∀y)(R(a,y) → T(a,y)) → ((∀y)R(a,y) → (∀y)T(a,y)) Axiom 4
7. (∀y)R(a,y) → (∀y)T(a,y)   5,6, modus ponens
8. (∀y)R(a,y)   2,tautology A∧B→B,modus ponens

9. $(\forall y)T(a,y)$ 7,8, modus ponens

10. $M(a) \land (\forall y)T(a,y)$ $A \land B$ can be deduced from A, B

11. $(\exists x)(M(x) \land (\forall y)T(x,y))$ 10, Axiom 7, modus ponens

27. The argument is:

$(\forall x)(C(x) \rightarrow (\exists y)W(x,y)) \land (\forall x)(\forall y)(W(x,y) \rightarrow S(x,y)) \land C(m) \rightarrow (\exists y)S(m,y)$

A proof sequence is:

1. $(\forall x)(C(x) \rightarrow (\exists y)W(x,y))$ hypothesis
2. $C(m) \rightarrow (\exists y)W(m,y)$ 1, Axiom 5, modus ponens
3. $C(m)$ hypothesis
4. $(\exists y)W(m,y)$ 2,3, modus ponens
5. $(\forall x)(\forall y)(W(x,y) \rightarrow S(x,y))$ hypothesis
6. $(\forall y)(W(m,y) \rightarrow S(m,y))$ 5, Axiom 5, modus ponens
7. $W(m,a)$ 4, Axiom 6, modus ponens
8. $W(m,a) \rightarrow S(m,a)$ 6, Axiom 5, modus ponens
9. $S(m,a)$ 7,8, modus ponens
10. $(\exists y)S(m,y)$ 9, Axiom 7, modus ponens

28. $x + 1 = y - 1$. For every x, if $x + 1 = y - 1$ before the statement, then $x = y - 2$; when the value of x is increased by 1, $x = y - 1$.

29. $2x > y$. For every x, if $2x > y$ before the statement, then $x > y/2$; when the value of x is doubled, $x > y$.

Section 1.3

*1. a) Converse: Healthy plant growth implies sufficient water.
   Contrapositive: If there is not healthy plant growth, then there is not sufficient water.

   b) Converse: Increased availability of microcomputers implies further technological advances.
   Contrapositive: If there is not increased availability of microcomputers, then there are no further technological advances.

   c) Converse: If there is a modification of the program then errors will be introduced.
   Contrapositive: No modification of the program implies that errors will not be introduced.

   d) Converse: Good insulation or all windows are storm windows implies fuel savings.
   Contrapositive: Poor insulation and some windows not storm windows implies no fuel savings.

2. For example:
   a) a nonsquare rectangle   b) 0

c) a short, blue-eyed redhead   d) a redhead who is short

3. Let $x = 2m$, $y = 2n$, where m and n are integers. Then $x + y = 2m + 2n = 2(m + n)$, where $m + n$ is an integer, so $x + y$ is even.

4. Let $x = 2m$, $y = 2n$ for integers m and n, and assume that $x + y$ is odd. Then $x + y = 2m + 2n = 2k + 1$ for some integer k or $2(m + n - k) = 1$ where $m + n - k$ is an integer. This is a contradiction since 1 is not even.

*5. Let $x = 2m + 1$, $y = 2n + 1$, where m and n are integers. Then $x + y = (2m + 1) + (2n + 1) = 2m + 2n + 2 = 2(m + n + 1)$, where $m + n + 1$ is an integer, so $x + y$ is even.

6. For two consecutive integers, one is even and one is odd. The product of an even integer and an odd integer is even by Example 1.49.

7. Let n be an integer. Case 1: n is even. Then n^2 is even by Example 1.43 and $n + n^2$ is even by Exercise 3.
Case 2: n is odd. Then n^2 is odd by Example 1.49 and $n + n^2$ is even by Exercise 5.

8. The contrapositive is: if $x + 1 \leq 0$, then $x \leq 0$. If $x + 1 \leq 0$, then $x \leq -1 < 0$, so $x < 0$ and therefore $x \leq 0$.

*9. If $x < y$ then multiplying both sides of the inequality by the positive numbers x and y in turn gives $x^2 < xy$ and $xy < y^2$ and therefore $x^2 < xy < y^2$ or $x^2 < y^2$.
If $x^2 < y^2$ then
$y^2 - x^2 > 0$ (definition of <)
$(y + x)(y - x) > 0$ (factoring)
$(y + x) < 0$ and $(y - x) < 0$ (a positive number is the product
       or   of two negatives or two positives)
$(y + x) > 0$ and $(y - x) > 0$
But it cannot be that $(y + x) < 0$ because y and x are both positive, therefore $y - x > 0$ and $y > x$.

10. $n + (n + 1) + (n + 2) = 3n + 3 = 3(n + 1)$

*11. Let $x = 2n + 1$. Then $x^2 = (2n + 1)^2 = 4n^2 + 4n + 1 = 4n(n + 1) + 1$. But $n(n + 1)$ is even (Exercise 6), so $n(n + 1) = 2k$ for some integer k. Therefore $x^2 = 4(2k) + 1 = 8k + 1$

12. $n^3 - (n - 1)^3 = n^3 - [n^3 - 3n^2 + 3n - 1] = 3n^2 - 3n + 1 = 3n(n - 1) + 1$. Then $n(n - 1)$ is even by Exercise 6, and

21

$3n(n - 1)$ is even by Example 1.49, thus $3n(n - 1) + 1$ is odd.

13. Proof by contradiction: assume that $m^2 + n^2 = k^2$ where m and n are odd integers and k is an integer. By Exercise 11, $m^2 = 8k_1 + 1$ and $n^2 = 8k_2 + 1$ for integers k_1 and k_2. Therefore $(8k_1 + 1) + (8k_2 + 1) = k^2$, or $2[4k_1 + 4k_2 + 1] = k^2$. Then 2 divides k^2, so 2 divides k, hence 4 is a factor of k^2, which can be written as 4x. Therefore $2[4k_1 + 4k_2 + 1] = 4x$ or $4k_1 + 4k_2 + 1 = 2x$. This is a contradiction because $4k_1 + 4k_2 + 1$ is odd while 2x is even.

14. Following the proof of Example 1.51, $\sqrt{4} = p/q$, $4 = p^2/q^2$, $4q^2 = p^2$, and 4 divides p^2. But we cannot now conclude that 4 divides p; for example $4|36$ but $4 \nmid 6$. In Example 1.51, 2 divides p^2 implies 2 divides p because 2 is prime.

15. Assume that $\sqrt{3}$ is rational. Then $\sqrt{3} = p/q$ where p and q are integers, $q \neq 0$, and p and q have no common factors (other then ± 1). If $\sqrt{3} = p/q$ then $3 = p^2/q^2$ or $3q^2 = p^2$. Then 3 divides p^2 so 3 divides p. Thus 3 is a factor of p or 9 is a factor of p^2, and the equation $3q^2 = p^2$ can be written $3q^2 = 9x$ or $q^2 = 3x$. Then 3 divides q^2 so 3 divides q. Therefore 3 is a common factor of p and q, a contradiction.

16. Assume that $\sqrt[3]{2}$ is rational. Then $\sqrt[3]{2} = p/q$ where p and q are integers, $q \neq 0$, and p and q have no common factors (other than ± 1). If $\sqrt[3]{2} = p/q$ then $2 = p^3/q^3$ or $2q^3 = p^3$. Then 2 divides p^3 so 2 divides p. Thus 2 is a factor of p, or 8 is a factor of p^3, and the equation $2q^3 = p^3$ can be written $2q^3 = 8x$ or $q^3 = 4x$. Then 2 divides 4x so 2 divides q^3 or 2 divides q. Therefore 2 is a common factor of p and q, a contradiction.

*17. Proof: if x is even, then $x = 2n$, and $(2n)(2n + 1)(2n + 2) = 2[(n)(2n + 1)(2n + 2)]$, which is even. If x is odd, then $x = 2n + 1$ and $(2n + 1)(2n + 2)(2n + 3) = 2[(2n + 1)(n + 1)(2n + 3)]$, which is even.

18. Counterexample: $2 + 3 + 4 = 9$

19. Counterexample: $3 \cdot 9 = 27$

20. Proof: if x is even, then $x = 2n$ and $2n + (2n)^3 = 2n + 8n^3 = 2(n + 4n^3)$, which is even. If x is odd, then $x = 2n + 1$

22

and $(2n + 1) + (2n + 1)^3 = (2n + 1) + (8n^3 + 12n^2 + 6n + 1)$
$= 8n^3 + 12n^2 + 8n + 2 = 2(4n^3 + 6n^2 + 4n + 1)$, which is even.

21. Proof: Suppose $x + \frac{1}{x} < 2$. Then $\frac{x^2 + 1}{x} < 2$.

Multiplying both sides by $x > 0$, $x^2 + 1 < 2x$ or $(x - 1)^2 < 0$, a contradiction. Therefore $x + \frac{1}{x} \geq 2$.

Section 1.4

*1. P(1): $4 \cdot 1 - 2 = 2(1)^2$ or $2 = 2$ true

Assume P(k): $2 + 6 + 10 + \cdots + (4k - 2) = 2k^2$

Show P(k + 1): $2 + 6 + 10 + \cdots + [4(k + 1) - 2] = 2(k + 1)^2$

$2 + 6 + 10 + \cdots + [4(k + 1) - 2]$ left side of P(k + 1)

$= 2 + 6 + 10 + \cdots + (4k - 2) + [4(k + 1) - 2]$

$= 2k^2 + 4(k + 1) - 2$ using P(k)

$= 2k^2 + 4k + 2$

$= 2(k^2 + 2k + 1)$

$= 2(k + 1)^2$ right side of P(k + 1)

2. P(1): $2 = 1(1 + 1)$ true

Assume P(k): $2 + 4 + 6 + \cdots + 2k = k(k + 1)$

Show P(k + 1): $2 + 4 + 6 + \cdots + 2(k + 1) = (k + 1)((k + 1) + 1)$

$2 + 4 + 6 + \cdots + 2(k + 1)$ left side of P(k + 1)

$= 2 + 4 + 6 + \cdots + 2k + 2(k + 1)$

$= k(k + 1) + 2(k + 1)$ using P(k)

$= (k + 1)(k + 2)$ factoring

$= (k + 1)((k + 1) + 1)$ right side of P(k + 1)

*3. P(1): $1 = 1(2 \cdot 1 - 1)$ true

Assume P(k): $1 + 5 + 9 + \ldots + (4k - 3) = k(2k - 1)$

Show P(k + 1): $1 + 5 + 9 + \ldots + [4(k + 1) - 3] = (k + 1)(2(k + 1) - 1)$

$1 + 5 + 9 + \ldots + [4(k + 1) - 3]$ left side of P(k + 1)

$= 1 + 5 + 9 + \ldots + (4k - 3) + [4(k + 1) - 3]$

$= k(2k - 1) + 4(k + 1) - 3$ using P(k)

$= 2k^2 - k + 4k + 1$

$= 2k^2 + 3k + 1$

$\quad = (k + 1)(2k + 1)$

$\quad = (k + 1)(2(k + 1) - 1) \qquad$ right side of $P(k + 1)$

4. $P(1)$: $1 = 1(1 + 1)(1 + 2)/6 = 2 \cdot 3/6 \qquad$ true

   Assume $P(k)$: $1 + 3 + 6 + \ldots + \dfrac{k(k + 1)}{2} = k(k + 1)(k + 2)/6$

   Show $P(k + 1)$: $1 + 3 + 6 + \ldots + \dfrac{(k + 1)(k + 2)}{2} = $
   $(k + 1)(k + 2)(k + 3)/6$

   $\quad 1 + 3 + 6 + \ldots \dfrac{(k + 1)(k + 2)}{2} \qquad$ left side of $P(k + 1)$

   $\quad = 1 + 3 + 6 + \ldots + \dfrac{k(k + 1)}{2} + \dfrac{(k + 1)(k + 2)}{2}$

   $\quad = k(k + 1)(k + 2)/6 + \dfrac{(k + 1)(k + 2)}{2} \qquad$ using $P(k)$

   $\quad = (k + 1)(k + 2)[\dfrac{k}{6} + \dfrac{1}{2}] \qquad$ factoring

   $\quad = (k + 1)(k + 2)[\dfrac{k + 3}{6}]$

   $\quad = (k + 1)(k + 2)(k + 3)/6 \qquad$ right side of $P(k + 1)$

5. $P(1)$: $6 - 2 = 1(3 \cdot 1 + 1) \qquad$ true

   Assume $P(k)$: $4 + 10 + 16 + \ldots + (6k - 2) = k(3k + 1)$

   Show $P(k + 1)$: $4 + 10 + 16 + \ldots + [6(k + 1) - 2] = $
   $(k + 1)[3(k + 1) + 1]$

   $\quad 4 + 10 + 16 + \ldots + [6(k + 1) - 2] \qquad$ left side of $P(k + 1)$

   $\quad = 4 + 10 + 16 + \ldots + (6k - 2) + [6(k + 1) - 2]$

   $\quad = k(3k + 1) + 6(k + 1) - 2 \qquad$ using $P(k)$

   $\quad = 3k^2 + k + 6k + 4$

   $\quad = 3k^2 + 7k + 4$

   $\quad = (k + 1)(3k + 4)$

   $\quad = (k + 1)[3(k + 1) + 1] \qquad$ right side of $P(k + 1)$

6. $P(1)$: $5 = 5(1 + 1)/2 \qquad$ true

   Assume $P(k)$: $5 + 10 + 15 + \ldots + 5k = 5k(k + 1)/2$

   Show $P(k + 1)$: $5 + 10 + 15 + \ldots + 5(k + 1) = 5(k + 1)(k + 2)/2$

   $\quad 5 + 10 + 15 + \ldots + 5(k + 1) \qquad$ left side of $P(k + 1)$

   $\quad = 5 + 10 + 15 + \ldots + 5k + 5(k + 1)$

$$= 5k(k + 1)/2 + 5(k + 1) \qquad \text{using } P(k)$$
$$= 5(k + 1)[\tfrac{k}{2} + 1] \qquad \text{factoring}$$
$$= 5(k + 1)[\tfrac{k + 2}{2}] \qquad \text{right side of } P(k + 1)$$

or write $5 + 10 + 15 + \ldots + 5n$ as

$$5(1 + 2 + 3 + \ldots + n)$$

and use the result of Practice 1.55

7. $P(1)$: $\quad 1^2 = 1(1 + 1)(2 + 1)/6 \qquad$ true

Assume $P(k)$: $\quad 1^2 + 2^2 + \ldots + k^2 = k(k + 1)(2k + 1)/6$

Show $P(k + 1)$: $\quad 1^2 + 2^2 + \ldots + (k + 1)^2 =$

$(k + 1)(k + 2)(2(k + 1) + 1)/6$

$$1^2 + 2^2 + \ldots + (k + 1)^2 \qquad \text{left side of } P(k + 1)$$
$$= 1^2 + 2^2 + \ldots + k^2 + (k + 1)^2$$
$$= k(k + 1)(2k + 1)/6 + (k + 1)^2 \qquad \text{using } P(k)$$
$$= (k + 1)\left[\frac{k(2k + 1)}{6} + k + 1\right] \qquad \text{factoring}$$
$$= (k + 1)\left[\frac{2k^2 + k + 6k + 6}{6}\right]$$
$$= (k + 1)\left[\frac{2k^2 + 7k + 6}{6}\right]$$
$$= \frac{(k + 1)(k + 2)(2k + 3)}{6}$$
$$= (k + 1)(k + 2)(2(k + 1) + 1)/6 \qquad \text{right side of } P(k + 1)$$

8. $P(1)$: $\quad 1^3 = 1^2(1 + 1)^2/4 \qquad$ true

Assume $P(k)$: $\quad 1^3 + 2^3 + \ldots + k^3 = k^2(k + 1)^2/4$

Show $P(k + 1)$: $\quad 1^3 + 2^3 + \ldots + (k + 1)^3 = (k + 1)^2(k + 2)^2/4$

$$1^3 + 2^3 + \ldots + (k + 1)^3 \qquad \text{left side of } P(k + 1)$$
$$= 1^3 + 2^3 + \ldots + k^3 + (k + 1)^3$$
$$= k^2(k + 1)^2/4 + (k + 1)^3 \qquad \text{using } P(k)$$
$$= (k + 1)^2\left[\tfrac{k^2}{4} + k + 1\right] \qquad \text{factoring}$$
$$= (k + 1)^2\left[\tfrac{k^2 + 4k + 4}{4}\right]$$
$$= (k + 1)^2(k + 2)^2/4 \qquad \text{right side of } P(k + 1)$$

*9. P(1): $1^2 = 1(2 - 1)(2 + 1)/3$ true

Assume P(k): $1^2 + 3^2 + \ldots + (2k - 1)^2 = k(2k - 1)(2k + 1)/3$

Show P(k + 1): $1^2 + 3^2 + \ldots + [2(k + 1) - 1]^2 =$

$(k + 1)(2(k + 1) - 1)(2(k + 1) + 1)/3$

$1^2 + 3^2 + \ldots + [2(k + 1) - 1]^2$ left side of P(k + 1)

$= 1^2 + 3^2 + \ldots + (2k - 1)^2 + [2(k + 1) - 1]^2$

$= \dfrac{k(2k - 1)(2k + 1)}{3} + [2(k + 1) - 1]^2$ using P(k)

$= \dfrac{k(2k - 1)(2k + 1)}{3} + (2k + 1)^2$

$= (2k + 1)\left[\dfrac{k(2k - 1)}{3} + 2k + 1\right]$ factoring

$= (2k + 1)\left[\dfrac{2k^2 - k + 6k + 3}{3}\right]$

$= \dfrac{(2k + 1)(2k^2 + 5k + 3)}{3}$

$= (2k + 1)(k + 1)(2k + 3)/3$

$= (k + 1)(2(k + 1) - 1)(2(k + 1) + 1)/3$ right side of P(k + 1)

10. P(1): $1^4 = 1(1 + 1)(2 + 1)(3 + 3 - 1)/30 = 2 \cdot 3 \cdot 5/30$ true

Assume P(k): $1^4 + 2^4 + \ldots + k^4 =$

$k(k + 1)(2k + 1)(3k^2 + 3k - 1)/30$

Show P(k + 1): $1^4 + 2^4 + \ldots + (k + 1)^4 =$

$(k + 1)(k + 2)(2(k + 1) + 1)(3(k + 1)^2 + 3(k + 1) - 1)/30$

$= (k + 1)(k + 2)(2k + 3)(3k^2 + 9k + 5)/30$

$= \dfrac{k + 1}{30}(2k^2 + 7k + 6)(3k^2 + 9k + 5)$

$= \dfrac{k + 1}{30}(6k^4 + 39k^3 + 91k^2 + 89k + 30)$

$1^4 + 2^4 + \ldots + (k + 1)^4$ left side of P(k + 1)

$= 1^4 + 2^4 + \ldots + k^4 + (k + 1)^4$

$= k(k + 1)(2k + 1)(3k^2 + 3k - 1)/30 + (k + 1)^4$ using P(k)

$= (k + 1)\left[\dfrac{k(2k + 1)(3k^2 + 3k - 1)}{30} + (k + 1)^3\right]$ factoring

$= \dfrac{(k + 1)}{30}[(2k^2 + k)(3k^2 + 3k - 1) + 30(k + 1)^3]$

$= \dfrac{k + 1}{30}(6k^4 + 6k^3 - 2k^2 + 3k^3 + 3k^2 - k + 30(k^3 + 3k^2 + 3k + 1))$

$= \dfrac{k + 1}{30}(6k^4 + 39k^3 + 91k^2 + 89k + 30)$ right side of P(k + 1)

11. $P(1)$: $1^2 = 1^2(1+1)^2/4$ true

    Assume $P(k)$: $(1 + 2 + \ldots + k)^2 = k^2(k+1)^2/4$

    Show $P(k+1)$: $(1 + 2 + \ldots + (k+1))^2 = (k+1)^2(k+2)^2/4$

$(1 + 2 + \ldots + (k+1))^2$ left side of $P(k+1)$

$= ((1 + 2 + \ldots + k) + (k+1))^2$

$= (1 + 2 + \ldots + k)^2 + 2(1 + 2 + \ldots + k)(k+1) + (k+1)^2$

                                                            squaring a binomial

$= \dfrac{k^2(k+1)^2}{4} + 2\dfrac{k(k+1)}{2}(k+1) + (k+1)^2$ using $P(k)$ and

                                                             Practice 1.55

$= (k+1)^2[\dfrac{k^2}{4} + k + 1]$

$= (k+1)^2[\dfrac{k^2 + 4k + 4}{4}]$

$= (k+1)^2(k+2)^2/4$ right side of $P(k+1)$

12. $P(1)$: $a^0 = \dfrac{a^1 - 1}{a - 1}$ or $1 = \dfrac{a - 1}{a - 1}$ true

    Assume $P(k)$: $1 + a + \ldots + a^{k-1} = \dfrac{a^k - 1}{a - 1}$

    Show $P(k+1)$: $1 + a + \ldots + a^k = \dfrac{a^{k+1} - 1}{a - 1}$

$1 + a + \ldots + a^k$ left side of $P(k+1)$

$= 1 + a + \ldots + a^{k-1} + a^k$

$= \dfrac{a^k - 1}{a - 1} + a^k$ using $P(k)$

$= \dfrac{a^k - 1 + a^k(a - 1)}{a - 1}$

$= \dfrac{a^k - 1 + a^{k+1} - a^k}{a - 1}$

$= \dfrac{a^{k+1} - 1}{a - 1}$ right side of $P(k+1)$

*13. $P(1)$: $\dfrac{1}{1\cdot 2} = \dfrac{1}{1 + 1}$ true

    Assume $P(k)$: $\dfrac{1}{1\cdot 2} + \dfrac{1}{2\cdot 3} + \ldots + \dfrac{1}{k(k+1)} = \dfrac{k}{k+1}$

    Show $P(k+1)$: $\dfrac{1}{1\cdot 2} + \dfrac{1}{2\cdot 3} + \ldots + \dfrac{1}{(k+1)(k+2)} = \dfrac{k+1}{k+2}$

$\dfrac{1}{1\cdot 2} + \dfrac{1}{2\cdot 3} + \ldots + \dfrac{1}{(k+1)(k+2)}$ left side of $P(k+1)$

$$= \frac{1}{1\cdot 2} + \frac{1}{2\cdot 3} + \cdots + \frac{1}{k(k+1)} + \frac{1}{(k+1)(k+2)}$$

$$= \frac{k}{k+1} + \frac{1}{(k+1)(k+2)} \qquad \text{using } P(k)$$

$$= \frac{k(k+2)+1}{(k+1)(k+2)}$$

$$= \frac{k^2 + 2k + 1}{(k+1)(k+2)}$$

$$= \frac{(k+1)^2}{(k+1)(k+2)}$$

$$= \frac{k+1}{k+2} \qquad \text{right side of } P(k+1)$$

14. $P(1)$: $\frac{1}{1\cdot 3} = \frac{1}{2+1}$ true

Assume $P(k)$: $\frac{1}{1\cdot 3} + \frac{1}{3\cdot 5} + \cdots + \frac{1}{(2k-1)(2k+1)} = \frac{k}{2k+1}$

Show $P(k+1)$: $\frac{1}{1\cdot 3} + \frac{1}{3\cdot 5} + \cdots + \frac{1}{(2(k+1)-1)(2(k+1)+1)} = \frac{k+1}{2(k+1)+1}$

$\frac{1}{1\cdot 3} + \frac{1}{3\cdot 5} + \cdots + \frac{1}{(2(k+1)-1)(2(k+1)+1)}$ left side of $P(k+1)$

$$= \frac{1}{1\cdot 3} + \frac{1}{3\cdot 5} + \cdots + \frac{1}{(2k-1)(2k+1)} + \frac{1}{(2(k+1)-1)(2(k+1)+1)}$$

$$= \frac{k}{2k+1} + \frac{1}{(2k+1)(2k+3)} \qquad \text{using } P(k)$$

$$= \frac{k(2k+3)+1}{(2k+1)(2k+3)}$$

$$= \frac{2k^2 + 3k + 1}{(2k+1)(2k+3)}$$

$$= \frac{(k+1)(2k+1)}{(2k+1)(2k+3)}$$

$$= \frac{k+1}{2k+3} \qquad \text{right side of } P(k+1)$$

*15. $P(2)$: $2^2 > 2 + 1$ true

Assume $P(k)$: $k^2 > k + 1$

Show $P(k+1)$: $(k+1)^2 > k + 2$

$(k+1)^2 = k^2 + 2k + 1$

$\qquad > (k+1) + 2k + 1 \qquad \text{using } P(k)$

$\qquad = 3k + 2 > k + 2$

16. $P(7)$: $7^2 > 5 \cdot 7 + 10$ or $49 > 45$ true
    Assume $P(k)$: $k^2 > 5k + 10$
    Show $P(k + 1)$: $(k + 1)^2 > 5(k + 1) + 10 = 5k + 15$
    $$\begin{aligned}(k + 1)^2 &= k^2 + 2k + 1 \\ &> (5k + 10) + 2k + 1 \quad \text{using } P(k) \\ &= 7k + 11 = 6k + k + 11 \\ &> 6k + 17 \quad \text{since } k > 6 \\ &> 5k + 15 \end{aligned}$$

17. $P(4)$: $4! > 4^2$ or $1 \cdot 2 \cdot 3 \cdot 4 = 24 > 16$ true
    Assume $P(k)$: $k! > k^2$
    Show $P(k + 1)$: $(k + 1)! > (k + 1)^2$
    $$\begin{aligned}(k + 1)! &= k!(k + 1) \\ &> k^2(k + 1) \quad \text{using } P(k) \\ &> (k + 1)(k + 1) \quad \text{since } k \geq 4 \\ &= (k + 1)^2 \end{aligned}$$

*18. $P(4)$: $2^4 < 4!$ or $16 < 24$ true
    Assume $P(k)$: $2^k < k!$
    Show $P(k + 1)$: $2^{k+1} < (k + 1)!$
    $$\begin{aligned}2^{k+1} &= 2 \cdot 2^k \\ &< 2 \cdot k! \quad \text{using } P(k) \\ &< (k + 1)k! \quad \text{since } k \geq 4 \\ &= (k + 1)! \end{aligned}$$

19. $P(2)$: $(1 + x)^2 > 1 + x^2$ or $1 + 2x + x^2 > 1 + x^2$ true
    because $x > 0$ implies $2x > 0$
    Assume $P(k)$: $(1 + x)^k > 1 + x^k$
    Show $P(k + 1)$: $(1 + x)^{k+1} > 1 + x^{k+1}$
    $$\begin{aligned}(1 + x)^{k+1} &= (1 + x)^k(1 + x) \\ &> (1 + x^k)(1 + x) \quad \text{using } P(k) \\ &= 1 + x^k + x + x^{k+1} \\ &> 1 + x^{k+1} \quad \text{because } x^k + x > 0 \end{aligned}$$

20. $P(1)$: $\left(\dfrac{a}{b}\right)^2 < \dfrac{a}{b}$ or $\dfrac{a^2}{b^2} < \dfrac{a}{b}$

    From $a < b$ and $a > 0$, $b > 0$, we get $aa < ab$ and then $aab < abb$. Dividing by the positive numbers b and b^2, we get $a^2 < \dfrac{ab^2}{b}$ and $\dfrac{a^2}{b^2} < \dfrac{a}{b}$.

Assume $P(k)$: $\left(\dfrac{a}{b}\right)^{k+1} < \left(\dfrac{a}{b}\right)^{k}$

Show $P(k+1)$: $\left(\dfrac{a}{b}\right)^{k+2} < \left(\dfrac{a}{b}\right)^{k+1}$

$$\left(\dfrac{a}{b}\right)^{k+2} = \left(\dfrac{a}{b}\right)^{k+1}\left(\dfrac{a}{b}\right)$$
$$< \left(\dfrac{a}{b}\right)^{k}\left(\dfrac{a}{b}\right) \quad \text{using } P(k)$$
$$= \left(\dfrac{a}{b}\right)^{k+1}$$

21. $P(2)$: $1 + 2 < 2^2$ or $3 < 4$ true
    Assume $P(k)$: $1 + 2 + \ldots + k < k^2$
    Show $P(k+1)$: $1 + 2 + \ldots + (k+1) < (k+1)^2$
    $1 + 2 + \ldots + (k+1)$
    $= 1 + 2 + \ldots + k + (k+1)$
    $< k^2 + k + 1$ using $P(k)$
    $< k^2 + 2k + 1 = (k+1)^2$

22. a) $P(1)$: $1 + \dfrac{1}{2} < 2$ true

    Assume $P(k)$: $1 + \dfrac{1}{2} + \ldots + \dfrac{1}{2^k} < 2$

    Show $P(k+1)$: $1 + \dfrac{1}{2} + \ldots + \dfrac{1}{2^{k+1}} < 2$

    $1 + \dfrac{1}{2} + \ldots + \dfrac{1}{2^{k+1}}$

    $= 1 + \dfrac{1}{2} + \ldots + \dfrac{1}{2^k} + \dfrac{1}{2^{k+1}}$

    $< 2 + \dfrac{1}{2^{k+1}}$ using $P(k)$

    but $2 + \dfrac{1}{2^{k+1}}$ is not less than 2.

    b) $P(1)$: $1 + \dfrac{1}{2} = 2 - \dfrac{1}{2}$ true

    Assume $P(k)$: $1 + \dfrac{1}{2} + \ldots + \dfrac{1}{2^k} = 2 - \dfrac{1}{2^k}$

    Show $P(k+1)$: $1 + \dfrac{1}{2} + \ldots + \dfrac{1}{2^{k+1}} = 2 - \dfrac{1}{2^{k+1}}$

    $1 + \dfrac{1}{2} + \ldots + \dfrac{1}{2^{k+1}}$

$$= 1 + \frac{1}{2} + \ldots + \frac{1}{2^k} + \frac{1}{2^{k+1}}$$

$$= 2 - \frac{1}{2^k} + \frac{1}{2^{k+1}} \qquad \text{using } P(k)$$

$$= 2 - \frac{2}{2^{k+1}} + \frac{1}{2^{k+1}} = 2 - \frac{1}{2^{k+1}}$$

*23. $P(1)$: $2^3 - 1 = 8 - 1 = 7$ and $7|7$ true
Assume $P(k)$: $7|2^{3k} - 1$ so $2^{3k} - 1 = 7m$ or
$2^{3k} = 7m + 1$ for some integer m
Show $P(k + 1)$: $7|2^{3(k+1)} - 1$
$2^{3(k+1)} - 1 = 2^{3k+3} - 1 = 2^{3k} \cdot 2^3 - 1 = (7m + 1)2^3 - 1$ (using
$P(k)$)
$= 7(2^3 m) + 8 - 1 = 7(2^3 m + 1)$ where $2^3 m + 1$ is an integer, so
$7|2^{3(k+1)} - 1$

24. $P(1)$: $3^2 + 7 = 9 + 7 = 16$ and $8|16$ true
Assume $P(k)$: $8|3^{2k} + 7$ so $3^{2k} + 7 = 8m$ or $3^{2k} = 8m - 7$ for
some integer m
Show $P(k + 1)$: $8|3^{2(k+1)} + 7$
$3^{2(k+1)} + 7 = 3^{2k+2} + 7 = 3^{2k} \cdot 3^2 + 7 = (8m - 7)9 + 7$ (using
$P(k)$)
$= 8(9m) - 63 + 7 = 8(9m) + 56 = 8(9m + 7)$ where $9m + 7$ is an
integer, so $8|3^{2(k+1)} + 7$

25. $P(1)$: $7 - 2 = 5$ and $5|5$ true
Assume $P(k)$: $5|7^k - 2^k$ so $7^k - 2^k = 5m$ or $7^k = 5m + 2^k$ for
some integer m
Show $P(k + 1)$: $5|7^{k+1} - 2^{k+1}$
$7^{k+1} - 2^{k+1} = 7 \cdot 7^k - 2^{k+1} = 7(5m + 2^k) - 2^{k+1}$ (using $P(k)$)
$= 5(7m) + 2^k (7 - 2) = 5(7m + 2^k)$ where $7m + 2^k$ is an
integer, so $5|7^{k+1} - 2^{k+1}$

26. $P(1)$: $13 - 6 = 7$ and $7|7$ true
Assume $P(k)$: $7|13^k - 6^k$ so $13^k - 6^k = 7m$ or
$13^k = 7m + 6^k$ for some integer m
Show $P(k + 1)$: $7|13^{k+1} - 6^{k+1}$
$13^{k+1} - 6^{k+1} = 13(13^k) - 6^{k+1} = 13(7m + 6^k) - 6^{k+1}$ (using $P(k)$)
$= 7(13m) + 6^k (13 - 6) = 7(13m + 6^k)$ where $13m + 6^k$ is an
integer, so $7|13^{k+1} - 6^{k+1}$

27. $P(1)$: $2 + (-1)^2 = 2 + 1 = 3$ and $3|3$ true
Assume $P(k)$: $3|2^k + (-1)^{k+1}$ so $2^k + (-1)^{k+1} = 3m$ or
$2^k = 3m - (-1)^{k+1}$ for some integer m
Show $P(k+1)$: $3|2^{k+1} + (-1)^{k+2}$
$2^{k+1} + (-1)^{k+2} = 2\cdot 2^k + (-1)^{k+2} = 2(3m - (-1)^{k+1}) + (-1)^{k+2}$
(using $P(k)$) $= 3(2m) - 2(-1)^{k+1} + (-1)^{k+2} =$
$3(2m) + (-1)^{k+1}(-2 + (-1)) = 3(2m) + (-1)^{k+1}(-3) =$
$3(2m - (-1)^{k+1})$ where $2m - (-1)^{k+1}$ is an integer,
so $3|2^{k+1} + (-1)^{k+2}$

28. $P(1)$: $2^{5+1} + 5^{1+2} = 2^6 + 5^3 = 64 + 125 = 189 = 27\cdot 7$
and $27|27\cdot 7$ true
Assume $P(k)$: $27|2^{5k+1} + 5^{k+2}$ so $2^{5k+1} + 5^{k+2} = 27m$
or $2^{5k+1} = 27m - 5^{k+2}$ for some integer m
Show $P(k+1)$: $27|2^{5(k+1)+1} + 5^{(k+1)+2}$
$2^{5(k+1)+1} + 5^{(k+1)+2} = 2^{5k+1+5} + 5^{k+3} = 2^{5k+1}\cdot 2^5 + 5^{k+3}$
$= (27m - 5^{k+2})2^5 + 5^{k+3}$ (using $P(k)$)
$= 27(m2^5) - 5^{k+2}\cdot 2^5 + 5^{k+2}\cdot 5 = 27(m2^5) + 5^{k+2}(5 - 2^5)$
$= 27(m2^5) + 5^{k+2}(-27) = 27(m2^5 - 5^{k+2})$ where
$m2^5 - 5^{k+2}$ is an integer, so $27|2^{5(k+1)+1} + 5^{(k+1)+2}$

29. $P(1)$: $3^{4+2} + 5^{2+1} = 3^6 + 5^3 = 729 + 125 = 854 = 61\cdot 14$
and $14|61\cdot 14$ true
Assume $P(k)$: $14|3^{4k+2} + 5^{2k+1}$ so $3^{4k+2} + 5^{2k+1} = 14m$ or
$3^{4k+2} = 14m - 5^{2k+1}$ for some integer m
Show $P(k+1)$: $14|3^{4(k+1)+2} + 5^{2(k+1)+1}$
$3^{4(k+1)+2} + 5^{2(k+1)+1} = 3^{4k+2}\cdot 3^4 + 5^{2k+3}$
$= (14m - 5^{2k+1})3^4 + 5^{2k+3}$ (using $P(k)$)
$= 14(m3^4) - 5^{2k+1}\cdot 3^4 + 5^{2k+1}\cdot 5^2 = 14(m3^4) - 5^{2k+1}(3^4 - 5^2)$
$= 14(m3^4) - 5^{2k+1}(81 - 25) = 14(m3^4) - 5^{2k+1}(56)$
$= 14(m3^4) - 5^{2k+1}(56) = 14(m3^4 - 4\cdot 5^{2k+1})$ where $m3^4 - 4\cdot 5^{2k+1}$
is an integer, so $14|3^{4(k+1)+2} + 5^{2(k+1)+1}$

30. $P(1)$: $7^2 + 16 - 1 = 49 + 15 = 64$ and $64|64$ true
Assume $P(k)$: $64|7^{2k} + 16k - 1$ so $7^{2k} + 16k - 1 = 64m$ or
$7^{2k} = 64m - 16k + 1$ for some integer m
Show $P(k+1)$: $64|7^{2(k+1)} + 16(k+1) - 1$
$7^{2(k+1)} + 16(k+1) - 1 = 7^{2k}\cdot 7^2 + 16(k+1) - 1$

$= (64m - 16k + 1) \cdot 7^2 + 16(k + 1) - 1$ (using P(k))
$= 64(m \cdot 7^2) - 16k \cdot 7^2 + 16k + 7^2 + 16 - 1$
$= 64(m \cdot 7^2) + 16k(1 - 7^2) + 64 = 64(m \cdot 7^2 + 1) + 16k(-48)$
$= 64(m \cdot 7^2 + 1 - 12k)$ where $m \cdot 7^2 + 1 - 12k$ is an integer,
so $64 | 7^{2(k+1)} + 16(k+1) - 1$

*31. P(1): $10 + 3 \cdot 4^3 + 5 = 10 + 192 + 5 = 207 = 9 \cdot 23$ true
Assume P(k): $9 | 10^k + 3 \cdot 4^{k+2} + 5$ so $10^k + 3 \cdot 4^{k+2} + 5 = 9m$
or $10^k = 9m - 3 \cdot 4^{k+2} - 5$ for some integer m
Show P(k + 1): $9 | 10^{k+1} + 3 \cdot 4^{k+3} + 5$
$10^{k+1} + 3 \cdot 4^{k+3} + 5 = 10 \cdot 10^k + 3 \cdot 4^{k+3} + 5$
$= 10(9m - 3 \cdot 4^{k+2} - 5) + 3 \cdot 4^{k+3} + 5$ (using P(k))
$= 9(10m) - 30 \cdot 4^{k+2} - 50 + 3 \cdot 4^{k+2} \cdot 4 + 5$
$= 9(10m) - 45 - 3 \cdot 4^{k+2}(10 - 4) = 9(10m - 5) - 18 \cdot 4^{k+2}$
$= 9(10m - 5 - 2 \cdot 4^{k+2})$ where $10m - 5 - 2 \cdot 4^{k+2}$ is an integer,
so $9 | 10^{k+1} + 3 \cdot 4^{k+3} + 5$

32. P(1): $x^1 - 1$ is divisible by $x - 1$ true
Assume P(k): $x^k - 1$ is divisible by $x - 1$, so $x^k - 1$
$= (x - 1)q(x)$ for some polynomial $q(x)$, or $x^k = (x - 1)q(x) + 1$
Show P(k + 1): $x^{k+1} - 1$ is divisible by $x - 1$
$x^{k+1} - 1 = x \cdot x^k - 1 = x((x - 1)q(x) + 1) - 1$
$= (x - 1)(xq(x)) + x - 1 = (x - 1)(xq(x) + 1)$ where $xq(x) + 1$
is a polynomial, so $x - 1$ divides $x^{k+1} - 1$

33. The statement to be proved is that $n(n + 1)(n + 2)$ is
divisible by 3 for $n \geq 1$.
P(1): $1(1 + 1)(1 + 2) = 6$ is divisible by 3   true
Assume P(k): $k(k + 1)(k + 2) = 3m$ for some integer m.
Show P(k + 1): $(k + 1)(k + 2)(k + 3)$ is divisible by 3
$(k + 1)(k + 2)(k + 3) = (k + 1)(k + 2)k + (k + 1)(k + 2)3$
$= 3m + (k + 1)(k + 2)3 = 3(m + (k + 1)(k + 2))$

34. Let n be arbitrary and show $x^n \cdot x^m = x^{n+m}$ for all $m \geq 1$.
Let $x^j \cdot x = x^{j+1}$ be equation (1).
m = 1: $x^n \cdot x^1 = x^{n+1}$ by (1)
Assume $x^n \cdot x^k = x^{n+k}$
Then $x^n \cdot x^{k+1} = x^n \cdot x^k \cdot x$ by (1)

$= x^{n+k} \cdot x$ by the inductive assumption
$= x^{n+k+1}$ by (1)

35. P(1): $A \wedge (B_1 \vee B_2) \leftrightarrow (A \wedge B_1) \vee (A \wedge B_2)$ is a tautology by 3b in the list of tautologies, Section 1.1
Assume $A \wedge (B_1 \vee B_2 \vee \ldots \vee B_k) \leftrightarrow (A \wedge B_1) \vee \ldots \vee (A \wedge B_k)$ is a tautology.
Then $A \wedge (B_1 \vee \ldots \vee B_k \vee B_{k+1}) \leftrightarrow A \wedge (B_1 \vee \ldots \vee B_k) \vee (A \wedge B_{k+1})$ is a tautology by 3b in the list of tautologies, Section 1.1, and $A \wedge (B_1 \vee \ldots \vee B_k) \leftrightarrow (A \wedge B_1) \vee \ldots \vee (A \wedge B_k)$ is a tautology by the inductive assumption, so by substitution of equivalent statements,
$A \wedge (B_1 \vee \ldots \vee B_k \vee B_{k+1}) \leftrightarrow (A \wedge B_1) \vee \ldots \vee (A \wedge B_k) \vee (A \wedge B_{k+1})$ is a tautology.

36. P(1): $(B_1 \wedge B_2)' \leftrightarrow B_1' \vee B_2'$ is a tautology by DeMorgan's Laws
Assume $(B_1 \wedge B_2 \wedge \ldots \wedge B_k)' \leftrightarrow B_1' \vee B_2' \vee \ldots \vee B_k'$ is a tautology
Then $(B_1 \wedge \ldots \wedge B_k \wedge B_{k+1})' \leftrightarrow (B_1 \wedge \ldots \wedge B_k)' \vee B_{k+1}'$ is a tautology by DeMorgan's Laws and $(B_1 \wedge \ldots \wedge B_k)' \leftrightarrow B_1' \vee \ldots \vee B_k'$ is a tautology by the inductive assumption, so by substitution of equivalent statements,
$(B_1 \wedge \ldots \wedge B_k \wedge B_{k+1})' \leftrightarrow B_1' \vee \ldots \vee B_k' \vee B_{k+1}'$ is a tautology

37. Proof is by induction on the length of the string. For a string of two characters, the single processed character could be 0 or 1. If it is 0, the parity bit stays 0, and the total number of 1's is zero, an even number. If it is 1, the parity bit switches from 0 to 1, and the total number of 1's is two, an even number. Now assume that a string of k characters contains an even number of 1's. A k + 1-length string is a k-length string with one new processed character. There are four cases:

	New character	Old parity bit	New parity bit
1.	0	0	0
2.	0	1	1
3.	1	0	1
4.	1	1	0

In cases (1) and (2) no 1's are added, so the total number of

1's remains the same as before, an even number. In case (3), the total number of 1's is increased by two, so is still an even number. In case (4), a 1 is added and a 1 is removed, so the total number of 1's remains the same as before, as even number.

38. $P(1)$ is $1 = 1 + 1$ which is not true.
39. $P(k) \to P(k + 1)$ fails when $k = 1$.
40. Let $T = \{t | P(t) \text{ is not true}\}$, and assume that $T \neq \phi$. By the Well-Ordering Principle, T has a smallest member t_0. Then $P(t_0)$ is not true, but $P(r)$ is true for all $r < t_0$, or all $r \leq t_0 - 1$. This contradicts implication (ii), so $T = \phi$.
41. $P(2)$: 2 is a prime number — true

    Assume that for all $r \leq k$, r is a prime number or a product of two or more prime numbers.

    Consider $k + 1$. If $k + 1$ is prime, we are done. If $k + 1$ is not prime, it can be written as $k + 1 = ab$ where $a \neq 1, b \neq 1$. Then $1 < a < k + 1$ and $1 < b < k + 1$, or $1 < a \leq k, 1 < b \leq k$. By the inductive hypothesis, a and b are either prime or the product of two or more primes. Thus $k + 1$ is the product of two or more primes.

42. $P(0)$: $j_0 = i_0^2$ true since $i = 1, j = 1$ before loop is entered
    Assume $P(k)$: $j_k = i_k^2$
    Show $P(k + 1)$: $j_{k+1} = (i_{k+1})^2$
    $j_{k+1} = j_k + 2i_k + 1 = i_k^2 + 2i_k + 1 = (i_k + 1)^2 = (i_{k+1})^2$
    At loop termination, $i = x$ and $j = i^2 = x^2$

*43. $P(0)$: $j_0 = (i_0 - 1)!$ true since $j = 1, i = 2$ before loop is entered and $1 = 1!$
    Assume $P(k)$: $j_k = (i_k - 1)!$
    Show $P(k + 1)$: $j_{k+1} = (i_{k+1} - 1)!$
    $j_{k+1} = j_k \cdot i_k = (i_k - 1)! i_k = (i_k)! = (i_{k+1} - 1)!$
    At loop termination, $i = x + 1$ and $j = x!$

44. $P(0)$: $j_0 = x^{i_0}$ true since $j = x, i = 1$ before loop is entered
    Assume $P(k)$: $j_k = x^{i_k}$
    Show $P(k + 1)$: $j_{k+1} = x^{i_{k+1}}$

$$j_{k+1} = j_k \cdot x = x^{i_k} \cdot x = x^{i_k+1} = x^{i_{k+1}}$$

At loop termination, $i = y$ and $j = x^y$

45. $P(0)$: $x = q_0 y + r_0$ true since $q = 0$, $r = x$ before loop is entered

    Assume $P(k)$: $x = q_k y + r_k$

    Show $P(k+1)$: $x = q_{k+1} y + r_{k+1}$

    $q_{k+1} y + r_{k+1} = (q_k + 1)y + r_k - y = q_k y + r_k = x$

    At loop termination, $x = qy + r$ and $0 \leq r < y$

Section 1.5

*1. 10, 20, 30, 40, 50
2. 2, $\frac{1}{2}$, 2, $\frac{1}{2}$, 2
3. 1, 5, 14, 30, 55
4. 1, $1 + \frac{1}{2}$, $1 + \frac{1}{2} + \frac{1}{3}$, $1 + \frac{1}{2} + \frac{1}{3} + \frac{1}{4}$, $1 + \frac{1}{2} + \frac{1}{3} + \frac{1}{4} + \frac{1}{5}$
*5. 1, 1, 2, 3, 5
6. 2, 2, 6, 14, 34
7. (b) and (c)
8. (a), (b), (c), (f), and (g)
*9. (a), (b), and (c)
10. (a), (b), and (d)
*11. a) $A(1) = 50{,}000$
    $A(n) = 3A(n-1)$ for $n \geq 2$
    b) 4
12. a) $P(1) = 500$
    $P(n) = (1.1)P(n-1)$ for $n \geq 2$
    b) 4
13. Algorithm $S(n)$
    if $n = 1$
    then $S \leftarrow 1$
    else begin
      $S \leftarrow 3 * S(n-1)$
    end
14. Algorithm $S(n)$
    if $n = 1$

```
        then S ← 2
        else begin
            S ← ½*S(n - 1)
        end
```
*15. Algorithm S(n)
```
    if n = 1
    then S ← 1
    else begin
        S ← S(n - 1) + (n - 1)
    end
```
16. Algorithm S(n)
```
    if n = 1
    then S ← 2
    else begin
        S ← (S(n - 1))²
    end
```
17. Algorithm S(n)
```
    if n = 1
    then S ← a
    else if n = 2
        then S ← b
    else begin
        S ← S(n - 2) + S(n - 1)
    end
```
18. Algorithm S(n)
```
    if n = 1
    then S ← p
    else begin
        S ← S(n - 1) + (-1)^(n-1) (n - 1)q
    end
```
*19. $S(n) = S(n - 1) + 5$
$\quad\quad = (S(n - 2) + 5) + 5 = S(n - 2) + 2*5$
$\quad\quad = (S(n - 3) + 5) + 2*5 = S(n - 3) + 3*5$
In general, $S(n) = S(n - k) + k*5$.
When $n - k = 1$, $k = n - 1$,
$S(n) = S(1) + (n - 1)*5 = 5 + (n - 1)*5 = n*5$
Now prove by induction that $S(n) = n*5$.

S(1): S(1) = 1*5 = 5   true
Assume S(k): S(k) = k*5
Show S(k + 1): S(k + 1) = (k + 1)*5
S(k + 1) = S(k) + 5 = k*5 + 5 = (k + 1)*5

20. $F(n) = nF(n - 1)$
    $= n((n - 1)F(n - 2)) = n(n - 1)F(n - 2)$
    $= n(n - 1)((n - 2)F(n - 3)) = n(n - 1)(n - 2)F(n - 3)$
    In general, $F(n) = n(n - 1)(n - 2)...(n - (k - 1))F(n - k)$
    When $n - k = 1$, $k = n - 1$,
    $F(n) = n(n - 1)(n - 2)...(2)F(1) = n(n - 1)(n - 2)...(2)(1) = n!$
    Now prove by induction that $F(n) = n!$
    F(1): $F(1) = 1! = 1$ true
    Assume F(k): $F(k) = k!$
    Show F(k + 1): $F(k + 1) = (k + 1)!$
    $F(k + 1) = (k + 1)F(k) = (k + 1)k! = (k + 1)!$

21. $T(n) = 2T(n - 1) + 1$
    $= 2(2T(n - 2) + 1) + 1 = 2^2 T(n - 2) + 1 + 2$
    $= 2^2(2T(n - 3) + 1) + 1 + 2 = 2^3 T(n - 3) + 1 + 2 + 2^2$
    In general, $T(n) = 2^k T(n - k) + 1 + 2 +...+ 2^{k-1}$
    When $n - k = 1$, $k = n - 1$,
    $T(n) = 2^{n-1} T(1) + 1 + 2 + ...+ 2^{n-2}$
    $= 1 + 2 + ... + 2^{n-2} + 2^{n-1}$
    $= 2^n - 1$ by Example 1.54
    Now prove by induction that $T(n) = 2^n - 1$
    T(1): $T(1) = 2^1 - 1 = 1$ true
    Assume T(k): $T(k) = 2^k - 1$
    Show T(k + 1): $T(k + 1) = 2^{k+1} - 1$
    $T(k + 1) = 2T(k) + 1 = 2(2^k - 1) + 1 = 2^{k+1} - 1$

22. $S(n) = S(n - 1) + 2n - 1$
    $= (S(n - 2) + 2(n - 1) - 1) + 2n - 1$
    $= S(n - 2) + 2 \cdot 2n - 1 - 3$
    $= (S(n - 3) + 2(n - 2) - 1) + 2 \cdot 2n - 1 - 3$
    $= S(n - 3) + 3 \cdot 2n - 1 - 3 - 5$
    In general, $S(n) = S(n - k) + k \cdot 2n - (1 + 3 + ... + (2k - 1))$
    $= S(n - k) + k \cdot 2n - k^2$ by Example 1.53
    When $n - k = 1$, $k = n - 1$,

$$S(n) = S(1) + (n-1) \cdot 2n - (n-1)^2$$
$$= 1 + 2n^2 - 2n - (n^2 - 2n + 1) = n^2$$

Now prove by induction that $S(n) = n^2$

$S(1)$: $S(1) = 1^2 = 1$ true

Assume $S(k)$: $S(k) = k^2$

Show $S(k+1)$: $S(k+1) = (k+1)^2$

$S(k+1) = S(k) + 2(k+1) - 1 = k^2 + 2k + 1 = (k+1)^2$

23. $T(n) = 2T\left(\frac{n}{2}\right) + n$

$$= 2\left(2T\left(\frac{n}{4}\right) + \frac{n}{2}\right) + n = 2^2 T\left(\frac{n}{4}\right) + 2n$$

$$= 2^2\left(2T\left(\frac{n}{8}\right) + \frac{n}{4}\right) + 2n = 2^3 T\left(\frac{n}{8}\right) + 3n$$

In general, $T(n) = 2^k T\left(\frac{n}{2^k}\right) + kn$

When $2^k = n$, $k = \log n$

$T(n) = nT(1) + n \log n$

$= n + n \log n = n(1 + \log n)$

Now prove by induction that $T(n) = n(1 + \log n)$ for $n = 2^m$

$T(1)$: $T(1) = 1(1 + \log 1) = 1(1 + 0) = 1$ true

Assume $T(k)$: $T(k) = k(1 + \log k)$ for $k = 2^{m-1}$

Show $T(2k)$: $T(2k) = 2k(1 + \log 2k)$

$T(2k) = 2T(k) + 2k = 2(k(1 + \log k)) + 2k$

$\qquad = 2k(\log 2 + \log k) + 2k = 2k \log 2k + 2k$

$\qquad = 2k(1 + \log 2k)$

24. $P(n) = 2P\left(\frac{n}{2}\right) + n^2$

$$= 2\left(2P\left(\frac{n}{4}\right) + \left(\frac{n}{2}\right)^2\right) + n^2 = 2^2 P\left(\frac{n}{4}\right) + n^2 + \frac{n^2}{2}$$

$$= 2^2\left(2P\left(\frac{n}{8}\right) + \left(\frac{n}{4}\right)^2\right) + n^2 + \frac{n^2}{2}$$

$$= 2^3 P\left(\frac{n}{8}\right) + n^2 + \frac{n^2}{2} + \frac{n^2}{4}$$

In general, $P(n) = 2^k P\left(\frac{n}{2^k}\right) + n^2\left(1 + \frac{1}{2} + \ldots + \frac{1}{2^{k-1}}\right)$

$$= 2^k P\left(\frac{n}{2^k}\right) + n^2\left(2 - \frac{1}{2^{k-1}}\right) \text{ by Exercise 22b of Section 1.4}$$

When $2^k = n$, $k = \log n$
$$P(n) = nP(1) + n^2(2 - \frac{1}{n/2})$$
$$= n + n^2(2 - \frac{2}{n}) = 2n^2 - n$$

Now prove by induction that $P(n) = 2n^2 - n$
$P(1)$: $P(1) = 2(1)^2 - 1 = 1$ true
Assume $P(k)$: $P(k) = 2k^2 - k$ for $k = 2^{m-1}$
Show $P(2k)$: $P(2k) = 2(2k)^2 - 2k$
$P(2k) = 2P(k) + (2k)^2 = 2(2k^2 - k) + (2k)^2 = 4k^2 - 2k + 4k^2$
        $= 2(2k)^2 - 2k$

CHAPTER 2

Section 2.1

*1. a) T b) F c) F d) F
2. a) T b) F c) T d) F ($\sqrt{2} \notin Q$)
3. *a) {0, 1, 2, 3, 4}
   *b) {3, 7, 10}
   c) {Washington, Adams, Jefferson}
   d) ϕ
   e) {-1, 3}
   f) {-3, -2, -1, 0, 1, 2, 3}
4. a) $\{x \mid x \in N \text{ and } 1 \leq x \leq 5\}$
   b) $\{x \mid x \in N \text{ and } x \text{ is odd}\}$
   c) {x | x is one of the Three Wise Men}
   d) {x | x is a nonnegative integer written in binary form}
*5. If $A = \{x \mid x = 2^n \text{ for } n \text{ a positive integer}\}$, then $16 \in A$.
   But if $A = \{x \mid x = 2 + n(n - 1) \text{ for } n \text{ a positive integer}\}$, then $16 \notin A$.
6. a) T b) T c) F d) T e) T f) F g) F h) T
7. *a) F; $\{1\} \in S$ but $\{1\} \notin R$ *b) T
   *c) F: $\{1\} \in S$, not $1 \in S$ *d) F; 1 is not a set; the correct statement is $\{1\} \subseteq U$
   *e) T   *f) F; $1 \notin S$
   g) T   h) T
   i) T   j) F; $3 \notin U$ and $\pi \notin U$
   k) T   l) T
   m) T   n) F
8. *a) T *b) F *c) F *d) T *e) T *f) F g) F h) T i) F j) F
9. a) For a = 1, b = -2, c = -24, the quadratic equation is
      $x^2 - 2x - 24 = 0$ or $(x + 4)(x - 6) = 0$, with solutions 6 and -4. Each of these is an even integer between -100 and 100, so each belongs to E.
   b) Here Q = {6,-4}, but E = {-4,-2,0,2,4}, and $Q \not\subseteq E$.
10. $\mathcal{P}(S) = \{\phi, \{a\}, \{b\}, \{a, b\}\}$
11. $\mathcal{P}(S) = \{\phi, \{1\}, \{2\}, \{3\}, \{4\}, \{1, 2\}, \{1, 3\}, \{1, 4\}, \{2, 3\},$
        $\{2, 4\}, \{3, 4\}, \{1, 2, 3\}, \{1, 2, 4\}, \{1, 3, 4\},$
        $\{2, 3, 4\}, \{1, 2, 3, 4\}\}$; $2^4 = 16$ elements

12. a) $\mathcal{P}(S) = \{\phi,\{\phi\},\{\{\phi\}\},\{\{\phi\ \{\phi\}\}\},\{\phi,\{\phi\}\},\{\phi,\{\phi,\{\phi\}\}\},\{\{\phi\},$
$\{\phi,\{\phi\}\}\},\{\phi,\{\phi\},\{\phi,\{\phi\}\}\}\ \}$

b) $\mathcal{P}(\mathcal{P}(S)) = \{\phi,\{\phi\},\{\{a\}\},\{\{b\}\},\{\{a,b\}\},\{\phi,\{a\}\},\{\phi,\{b\}\},$
$\{\phi,\{a,b\}\},\{\{a\},\{b\}\},\{\{a\},\{a,b\}\},\{\{a\},\{a,b\}\},$
$\{\phi,\{a\},\{b\}\},\{\phi,\{a\},\{a,b\}\},\{\phi,\{b\},\{a,b\}\},$
$\{\{a\},\{b\},\{a,b\}\},\{\phi,\{a\},\{b\},\{a,b\}\}\ \}$

13. a) $x = 1, y = 5$
    b) $x = 8, y = 7$
    c) $x = 1, y = 4$

14. a) If $x = u$ and $y = v$, then clearly $\{\{x\},\{x,y\}\} = \{\{u\},\{u,v\}\}$. Now assume $\{\{x\},\{x,y\}\} = \{\{u\},\{u,v\}\}$. Then $\{x\} = \{u\}$ or $\{x\} = \{u,v\}$. If $\{x\} = \{u,v\}$, then $u = v = x$; also $\{u\} = \{x,y\}$, and $x = y = u$. Thus $x = u$ and $y = v$. If $\{x\} = \{u\}$, then $x = u$; also $\{x,y\} = \{u,v\}$ and, since $x = u, y = v$.

    b) For example, the ordered triples $(1,1,2)$ and $(1,2,1)$ expressed in set form would be
    $\{\{1\},\{1,1\},\{1,1,2\}\} = \{\{1\},\{1,2\}\}$ and
    $\{\{1\},\{1,2\},\{1,2,1\}\} = \{\{1\},\{1,2\}\}$, respectively, and distinct ordered triples would have the same representation.

15. *a) binary operation *b) no; $0 \circ 0 \notin N$
    *c) binary operation d) no; $\ln x$ undefined for $x \leq 0$
    e) unary operation   f) no; closure fails
    g) no; uniqueness fails (assuming there are two people in Arkansas of the same height) h) binary operation
    i) no; operation undefined for $x = 0$ j) binary operation

16. n^{n^2}; each entry in the n x n matrix has n possible answers. We multiply the number of possibilities. Thus there are $\underbrace{n \cdot n \cdots \cdot n}_{n} = n^n$ ways to complete each row, and $\underbrace{n^n \cdot n^n \cdots \cdot n^n}_{n} = n^{n^2}$ ways to complete the table.

	x_1 $\cdots$ x_n
x_1	
$\vdots$	
x_n	

17. a) $(A + B)*(C - D) = ((A + B)*(C - D)) \rightarrow AB+CD-*$
    b) $A**B - C*D = ((A**B) - (C*D)) \rightarrow AB**CD*-$
    c) $A*C + B/(C + D*B) = ((A*C) + (B/(C + (D*B)))) \rightarrow AC*BCDB*+/+$

42

18. a) {t} b) {p, q, r, s, t, u} c) {q, r, v, w} d) φ
    e) {r, v} f) {u, w}
    g) {(p, r), (p, t), (p, v), (q, r), (q, t), (q, v), (r, r), (r, t), (r, v), (s, r), (s, t), (s, v)}
    h) {q, r, v}

19. *a) {1, 2, 4, 5, 6, 8, 9}   *b) {4, 5}
    *c) {2, 4}   d) {1, 2, 3, 4, 5, 9}
    e) {2, 6, 8}   f) {0, 1, 3, 7, 9}
    g) φ   *h) {0, 1, 2, 3, 6, 7, 8, 9}
    i) {2, 3}   j) {0, 1, 3, 4, 7, 9}
    k) {2, 6, 8}   l) {2, 3}
    m) {(1, 2), (1, 3), (1, 4), (4, 2), (4, 3), (4, 4), (5, 2), (5, 3), (5, 4), (9, 2), (9, 3), (9, 4)}

20. a) B'   b) B ∩ C
    c) A ∩ B   d) B' ∩ C
    e) B' ∩ C' (or (B ∪ C)' or B' - C)

21. *a) T b) T *c) F (Let A = {1, 2, 3}, B = {1, 3, 5}, S = {1, 2, 3, 4, 5}. Then (A ∩ B)' = {2, 4, 5} but A' ∩ B' = {4, 5} ∩ {2, 4} = {4}.) d) T *e) F (Take A, B, and S as in (c), then A - B = {2}, (B - A)' = {1, 2, 3, 4}.) f) T g) F h) F (Order matters in ordered pairs, so if A = {1, 2}, B = {3, 4}, then B x A = {(3, 1), (3, 2), (4, 1), (4, 2)} and A x B = {(1, 3), (1, 4), (2, 3), (2, 4)}) i) T j) T

22.* a) B ⊆ A   b) A ⊆ B
     c) A = φ   d) B ⊆ A
     e) A = B

23. a)

    b) {2, 4, 6, 7, 9}
    c) x ε (A ∪ B) - (A ∩ B) ⟷ x ε (A ∪ B) and x ε (A ∩ B)'
       ⟶ (xεA or xεB) and x∉A ∩ B
       ⟵ (xεA and x∉A ∩ B) or (xεB and x∉A ∩ B)

43

$\longleftrightarrow$ ($x \in A$ and $x \notin B$) or ($x \in B$ and $x \notin A$)
$\longleftrightarrow$ $x \in (A - B) \cup (B - A)$

d) ϕ; A

e) $A \oplus B = (A \cup B) - (A \cap B) = (B \cup A) - (B \cap A) = B \oplus A$

f) First note that $(A \oplus B)' = (A \cap B) \cup (A' \cap B')$.
Then $(A \oplus B) \oplus C = [(A \oplus B) - C] \cup [C - (A \oplus B)]$
$= [(A \oplus B) \cap C'] \cup [C \cap (A \oplus B)']$
$= [((A - B) \cup (B - A)) \cap C'] \cup [C \cap ((A \cap B) \cup (A' \cap B'))]$
$= (A \cap B' \cap C') \cup (B \cap A' \cap C') \cup (C \cap A \cap B) \cup (C \cap A' \cap B')$
$= (A \cap B \cap C) \cup (A \cap B' \cap C') \cup (B \cap C' \cap A') \cup (C \cap B' \cap A')$
$= A \cap [(B \cap C) \cup (B' \cap C')] \cup [(B \cap C') \cup (C \cap B')] \cap A'$
$= [A \cap (B \oplus C)'] \cup [(B \oplus C) \cap A']$
$= (A - (B \oplus C)) \cup ((B \oplus C) - A)$
$= A \oplus (B \oplus C)$

24. a) $A_1 \cup A_2 \cup \ldots \cup A_n = \{x \mid x \text{ belongs to some } A_i \text{ for } 1 \leq i \leq n\}$

b) $A_1 = A_1$ $n = 1$
$A_1 \cup A_2 = \{x \mid x \in A_1 \text{ or } x \in A_2\}$ $n = 2$
$A_1 \cup A_2 \cup \ldots \cup A_n = (A_1 \cup \ldots \cup A_{n-1}) \cup A_n$ $n > 2$

25. a) $|A| + |B|$ counts all elements in A and in B, but counts anything in $A \cap B$ twice, hence $|A \cap B|$ must be subtracted from $|A| + |B|$ to get $|A \cup B|$.

b) $|A \cup B \cup C| = |A \cup (B \cup C)| = |A| + |B \cup C|$
    $- |A \cap (B \cup C|$ (by part (a))
$= |A| + |B| + |C| - |B \cap C|$ (by part (a) and
    $- |(A \cap B) \cup (A \cap C)|$ set identity 3b)
$= |A| + |B| + |C| - |B \cap C|$
    $- (|A \cap B| + |A \cap C| - |A \cap B \cap C|)$
    (by part (a))
$= |A| + |B| + |C| - |A \cap B| - |A \cap C|$
    $- |B \cap C| + |A \cap B \cap C|$

c) Let A = {students who wrote programs that ran in ≤ 2 minutes}
B = {students who wrote programs that produced ≤ 15 pages of output}
C = {students who wrote programs with ≤ 5 variable names}
Then $|A \cup B \cup C| = 35$, $|A| = 19$, $|B| = 19$, $|C| = 15$, $|A \cap B| = 9$, $|B \cap C| = 7$, $|A \cap C| = 6$. Using the

equation from part (b),
$35 = 19 + 19 + 15 - 9 - 6 - 7 + |A \cap B \cap C|$, so
$|A \cap B \cap C| = 4$

26. (1a) $x \in A \cup B \leftrightarrow x \in A$ or $x \in B \leftrightarrow x \in B$ or
$x \in A \leftrightarrow x \in B \cup A$

(1b) $x \in A \cap B \leftrightarrow x \in A$ and $x \in B \leftrightarrow x \in B$ and
$x \in A \leftrightarrow x \in B \cap A$

(2a) $x \in (A \cup B) \cup C \leftrightarrow x \in (A \cup B)$ or $x \in C$
$\leftrightarrow (x \in A$ or $x \in B)$ or $x \in C \leftrightarrow x \in A$ or $x \in B$ or
$x \in C \leftrightarrow x \in A$ or $(x \in B$ or $x \in C) \leftrightarrow x \in A$ or
$x \in (B \cup C) \leftrightarrow x \in A \cup (B \cup C)$

(2b) $x \in (A \cap B) \cap C \leftrightarrow x \in (A \cap B)$ and $x \in C$
$\leftrightarrow (x \in A$ and $x \in B)$ and $x \in C \leftrightarrow x \in A$ and
$x \in B$ and $x \in C \leftrightarrow x \in A$ and $(x \in B$ and $x \in C)$
$\leftrightarrow x \in A$ and $x \in (B \cap C) \leftrightarrow x \in A \cap (B \cap C)$

(3b) $x \in A \cap (B \cup C) \leftrightarrow x \in A$ and $x \in (B \cup C) \leftrightarrow x \in A$
and $(x \in B$ or $x \in C) \leftrightarrow (x \in A$ and $x \in B)$ or $(x \in A$
and $x \in C) \leftrightarrow x \in A \cap B$ or $x \in A \cap C \leftrightarrow x \in (A \cap B)$
$\cup (A \cap C)$

(4b) $x \in A \cap S \rightarrow x \in A$ and $x \in S \rightarrow x \in A$
$x \in A \rightarrow x \in A$ and $x \in S$ because $A \subseteq S \rightarrow x \in A \cap S$

(5a) $x \in A \cup A' \rightarrow x \in A$ or $x \in A' \rightarrow x \in S$ or $x \in S$
because $A \subseteq S$, $A' \subseteq S \rightarrow x \in S$
$x \in S \rightarrow (x \in S$ and $x \in A)$ or $(x \in S$ and $x \notin A)$
$\rightarrow x \in A$ or $x \in A' \rightarrow x \in A \cup A'$

(5b) For any x such that $x \in A \cap A'$, it follows that $x \in A$
and $x \in A'$ or x belongs to A and x does not belong to
A. This is a contradiction, so no x belongs to $A \cap A'$,
and $A \cap A' = \phi$.

27. a) $x \in (A \cup B)' \leftrightarrow x \notin (A \cup B) \leftrightarrow x$ does not belong to
either A or B $\leftrightarrow x \notin A$ and $x \notin B \leftrightarrow x \in A'$ and
$x \in B' \leftrightarrow x \in A' \cap B'$.

b) $x \in (A \cap B)' \leftrightarrow x \notin A \cap B \leftrightarrow x$ does not belong to
both A and B $\leftrightarrow x \notin A$ or $x \notin B \leftrightarrow x \in A'$ or $x \in B'$
$\leftrightarrow x \in A' \cup B'$.

c) $x \in A \cup (B \cap A) \leftrightarrow x \in A$ or $x \in (B \cap A) \leftrightarrow$
$x \in A$ or $(x \in B$ and $x \in A) \leftrightarrow x \in A$

d) $x \in (A \cap B')' \cup B \leftrightarrow x \in (A \cap B')'$ or $x \in B$
   $\leftrightarrow x \in (A' \cup B)$ or $x \in B$ by part (b) and $(B')' = B$
   $\leftrightarrow x \in A'$ or $x \in B$ or $x \in B \leftrightarrow x \in A'$ or $x \in B$
   $\leftrightarrow x \in A' \cup B$

e) $x \in (A \cap B) \cup (A \cap B') \leftrightarrow x \in A \cap B$ or $x \in A \cap B'$
   $\leftrightarrow (x \in A$ and $x \in B)$ or $(x \in A$ and $x \in B')$
   $\leftrightarrow x \in A$

28. a) Proof is by induction on n
   For $n = 2$, $(A_1 \cup A_2)' = A_1' \cap A_2'$ by Exercise 27
   Assume that $(A_1 \cup \ldots \cup A_k)' = A_1' \cap \ldots \cap A_k'$.
   Then $(A_1 \cup \ldots \cup A_k \cup A_{k+1})' = ((A_1 \cup \ldots \cup A_k) \cup A_{k+1})'$
   $= (A_1 \cup \ldots \cup A_k)' \cap A_{k+1}'$ by Exercise 27
   $= (A_1' \cap \ldots \cap A_k') \cap A_{k+1}'$ by inductive hypothesis
   $= A_1' \cap \ldots \cap A_k' \cap A_{k+1}'$

   b) Proof is by induction on n
   For $n = 2$, $(A_1 \cap A_2)' = A_1' \cup A_2'$ by Exercise 27
   Assume that $(A_1 \cap \ldots \cap A_k)' = A_1' \cup \ldots \cup A_k'$.
   Then $(A_1 \cap \ldots \cap A_k \cap A_{k+1})' = ((A_1 \cap \ldots \cap A_k) \cap A_{k+1})'$
   $= (A_1 \cap \ldots \cap A_k)' \cup A_{k+1}'$ by Exercise 27
   $= (A_1' \cup \ldots \cup A_k') \cup A_{k+1}'$ by inductive hypothesis
   $= A_1' \cup \ldots \cup A_k' \cup A_{k+1}'$

29. a) *1) $(A \cup B) \cap (A \cup B') = A \cup (B \cap B')$ (3a)
   $= A \cup \phi$ (5b)
   $= A$ (4a)

   2) $[((A \cap C) \cap B) \cup ((A \cap C) \cap B')] \cup (A \cap C)'$
   $= [(A \cap C) \cap (B \cup B')] \cup (A \cap C)'$ (3b)
   $= [(A \cap C) \cap S] \cup (A \cap C)'$ (5a)
   $= (A \cap C) \cup (A \cap C)'$ (4b)
   $= S$ (5a)

   3) $(A \cup C) \cap [(A \cap B) \cup (C' \cap B)]$
   $= (A \cup C) \cap [(B \cap A) \cup (B \cap C')]$ (1b)
   $= (A \cup C) \cap [B \cap (A \cup C')]$ (3b)
   $= (A \cup C) \cap [(A \cup C') \cap B]$ (1b)
   $= [(A \cup C) \cap (A \cup C')] \cap B$ (2b)
   $= [A \cup (C \cap C')] \cap B$ (3a)

$$= (A \cup \phi) \cap B \tag{5b}$$
$$= A \cap B \tag{4a}$$

b) $(A \cap B) \cup (A \cap B') = A$
   $[((A \cup C) \cup B) \cap ((A \cup C) \cup B')] \cap (A \cup C)' = \phi$
   $(A \cap C) \cup [(A \cup B) \cap (C' \cup B)] = A \cup B$

30. a) $\bigcup_{i \in I} A_i = \{x \mid x \in (-1,1)\}$; $\bigcap_{i \in I} A_i = \{0\}$
    b) $\bigcup_{i \in I} A_i = \{x \mid x \in [-1,1]\}$; $\bigcap_{i \in I} A_i = \{0\}$

*31. An enumeration of the set is
    1, 3, 5, 7, 9, 11, ...

32. An enumeration of Z is
    0, 1. -1, 2, -2, 3, -3, 4, -4, 5, -5, ...

33. An enumeration of the set is
    a, aa, aaa, aaaa, ...

34. An enumeration of the set is shown by the arrow through the array

    (0, 0) (0, 1) (0, -1) (0, 2) (0, -2) (0, 3) (0, -3) ...
    (1, 0) (1, 1) (1, -1) (1, 2) (1, -2) (1, 3) (1, -3) ...
    (-1, 0) (-1, 1) (-1, -1) (-1, 2) (-1, -2) (-1, 3) (-1, -3)...
    (2, 0) (2, 1) (2, -1) (2, 2) (2, -2) (2, 3) (2, -3) ...
    .
    .
    .

35. Assume that the set has an enumeration
    $Z_{11}, Z_{12}, Z_{13}, Z_{14}, \ldots$
    $Z_{21}, Z_{22}, Z_{23}, Z_{24}, \ldots$
    $Z_{31}, Z_{32}, Z_{33}, Z_{34}, \ldots$
    .
    .
    .

    Now construct an infinite sequence Z of positive integers with $Z = Z_1, Z_2, Z_3, \ldots$ such that $Z_i \neq Z_{ii}$ for all i. Then Z differs from every sequence in the enumeration, yet is a member of the set. This is a contradiction, so the set is uncountable.

36. Let A be a countable set. Then A is finite or countably infinite. If A is finite and $B \subseteq A$, then B is finite, hence countable. If A is countably infinite, let $a_1, a_2, a_3, \ldots$ be an enumeration of A. Using this same list but eliminating elements in A - B gives an enumeration of B.

37. Let A and B be denumerable sets with enumerations
$A = a_1, a_2, a_3, \ldots$ and $B = b_1, b_2, b_3, \ldots$
Then use the list $a_1, b_1, a_2, b_2, a_3, b_3, \ldots$ and eliminate any duplicates. This will be an enumeration of $A \cup B$, which is therefore denumerable.

38. $B = \{S | S \text{ is a set and } S \notin S\}$. Then either $B \in B$ or $B \notin B$. If $B \in B$, then B has the property of all members of B, namely $B \notin B$. Hence both $B \in B$ and $B \notin B$ are true. If $B \notin B$, then B has the property characterizing members of B, hence $B \in B$. Therefore both $B \notin B$ and $B \in B$ are true.

Section 2.2

*1. $5 \cdot 3 \cdot 2 = 30$

*2. $4 \cdot 2 \cdot 2 = 16$

3. $4 \cdot 8 \cdot 6 = 192$

4. $4^{20} \cdot 5^{10}$

5. $26^3 \cdot 10^2$

6. $52^3 \cdot 10^2$

7. $45 \cdot 13 = 585$

8. 10^9

9. $26 \cdot 26 \cdot 26 \cdot 1 \cdot 1 = 17,576$

10. $2 \cdot 4 \cdot 4 = 32$

11. $26 + 26 \cdot 10 = 286$

12. $17 \cdot 16 + 24 \cdot 23 = 824$

*13. $2^8 = 256$

14. $2^6 = 64$

*15. $1 \cdot 2^7$ (begin with 0) $+ 1 \cdot 2^6 \cdot 1$ (begin with 1, end with 0)
$= 2^7 + 2^6 = 192$

16. $2 \cdot 1 \cdot 2^6 = 2^7 = 128$

17. $1 \cdot 1 \cdot 1 \cdot 2^5 = 32$

18. 8 (one for each digit at which the 0 occurs)

19. $1 \cdot 1 \cdot 2^6$ (begin with 10) + $1 \cdot 1 \cdot 1 \cdot 2^5$ (begin with 110) + $1 \cdot 1 \cdot 1 \cdot 2^5$ (begin with 010) + $1 \cdot 1 \cdot 1 \cdot 2^5$ (begin with 000)
    $= 2^6 + 3 \cdot 2^5 = 160$
20. $2 \cdot 2 \cdot 2 \cdot 2 \cdot 1 \cdot 1 \cdot 1 \cdot 1 = 2^4 = 16$
*21. $52 \cdot 52 = 2704$
22. $4 \cdot 4 = 16$
*23. $4 \cdot 4 = 16$ ways to get 2 of one kind; there are 13 distinct "kinds", so by the Addition Principle, the answer is
    $16 + 16 + \ldots + 16 = 13 \cdot 16 = 208$
24. $4 \cdot 48$ (flower king, bird nonking) + $4 \cdot 48$ (bird king, flower nonking) = 384
25. Face value of 5 can occur in 4 disjoint ways:

flower face value	bird face value
1	4
2	3
3	2
4	1

    Each has $4 \cdot 4$ ways of occurring, so the total is $4 \cdot 16 = 64$
26. Face value of less than 5 can occur in the disjoint ways shown below:

flower face value	bird face value
1	1, 2, or 3
2	1 or 2
3	1

    Each has $4 \cdot 4$ ways of occurring, so the total is $6 \cdot 16 = 96$
27. $40 \cdot 40 = 1600$
28. $12 \cdot 52$ (flower face card, any bird card) + $40 \cdot 12$ (flower non-face card, bird face card) = 1104
    or $52 \cdot 52$ (total number of hands - Exercise 21)
    $- 40 \cdot 40$ (hands with no face cards - Exercise 27) = 1104
29. For $m = 2$, the result follows from the Multiplication Principle.
    Assume that for $m = k$, there are $n_1 \cdots n_k$ possible outcomes for the sequence of events 1 to k.
    Let $m = k + 1$. Then the sequence of events 1 to $k + 1$ consists of the sequence of events 1 to k followed by event $k + 1$. The sequence of events 1 to k has $n_1 \cdots n_k$

possible outcomes by the inductive hypothesis. The sequence 1 to k followed by event k + 1 then has $(n_1 \cdots n_k)n_{k+1}$ outcomes by the Multiplication Principle, which equals $n_1 \cdots n_{k+1}$.

30. Let A and B denote the sets of outcomes of two events. If the events are disjoint, then $A \cap B = \phi$. Also, $|A|$ = the number of outcomes of one event, $|B|$ the number of outcomes of the other, and $A \cup B$ is the set of outcomes of either event. By the Addition Principle, $|A \cup B| = |A| + |B|$, which agrees with the Principle of Inclusion and Exclusion because $|A \cap B| = |\phi| = 0$.

Section 2.3

1. *a) 42 b) 6720 c) 360
   d) $\frac{n!}{(n-1)!} = n$ e) $\frac{n!}{(n-(n-1))!} = \frac{n!}{1!} = n!$

2. $9! = 362,880$

3. $14! \approx 87,178,291,000$

4. $8! = 40,320$; $3 \cdot 7! = 15,120$

*5. $\frac{5!\,(\text{total permutations})}{3!\,(\text{arrangement of the 3 R's for each distinguished permutation})}$
   $= 5 \cdot 4 = 20$

6. $\frac{6!\,(\text{total permutations})}{6\,(\text{each of 6 places in circle for arrangement to start})}$
   $= 5! = 120$

*7. $P(15,3) = \frac{15!}{12!} = 15 \cdot 14 \cdot 13 = 2730$

8. a) $(26)^3$
   b) $P(26,3) = 26 \cdot 25 \cdot 24 = 15600$

9. $(2!)(11!)(8!) = 2(39,916,800)(40,320)$

10. $11!$ (arrange men)$\cdot P(12,8)$ (fill 12 slots with 8 women)

11. *a) 120 b) 36
    c) 28 d) $\frac{n!}{(n-1)!\,1!} = n$

12. $C(n,n)$ is the number of ways to select n objects from n objects, which is 1; $C(n,n) = \frac{n!}{n!0!} = \frac{n!}{n! \cdot 1} = 1$. $C(n,1)$ is the number of ways to select 1 object from n objects, which is n; $C(n,1) = \frac{n!}{1!(n-1)!} = \frac{n!}{(n-1)!} = \frac{n(n-1)!}{(n-1)!} = n$

*13. $C(300,25) = \frac{300!}{25!275!}$

14. $C(18,11)$
15. $C(17,5) \cdot C(23,7) = (6188)(245,157)$
16. $C(21,4) \cdot C(11,3)$
17. $C(7,1) \cdot C(14,1) \cdot C(4,1) \cdot C(5,1) \cdot C(2,1) \cdot C(3,1) = 7 \cdot 14 \cdot 4 \cdot 5 \cdot 2 \cdot 3$
    $= 11,760$
18. $C(14,2) \cdot C(21,4)$ (2 from manufacturing, 4 from the others)
19. $C(3,1) \cdot C(30,5)$ (1 from marketing, 5 from non-accounting and non-marketing)
20. all committees - (none or 1 from manufacturing)
    $= C(35,6) - (C(21,6) + C(14,1) \cdot C(21,5))$
*21. $C(13,3) \cdot C(13,2)$
22. $C(13,5)$
*23. $C(13,5) + C(13,5) + C(13,5) + C(13,5) = 4C(13,5)$
24. $C(12,5)$
25. exactly 4 of a kind + exactly 3 of a kind =
    $13 \cdot C(4,4) \cdot C(48,1) + 13 \cdot C(4,3) \cdot C(48,2)$
26. $13C(4,3) \cdot 12C(4,2)$ (3 of one kind, 2 of another kind)
*27. $C(12,4) = 495$
28. $C(5,2) \cdot C(7,2) = (10)(21) = 210$
29. $C(5,4) + C(7,4) = 5 + 35 = 40$
30. $C(7,3) \cdot C(5,1) + C(7,4) = 210$
31. $C(60,2)$
32. $C(60,1) + C(60,2)$
33. $C(59,7)$
34. $C(2,1) \cdot C(58,6)$
*35. $C(12,3) = 220$
36. all committees - no independent $= C(12,3) - C(8,3) = 164$
*37. no Democrats + no Republicans - all independents (so as not to count twice) $= C(7,3) + C(9,3) - C(4,3) = 115$
38. all committees - (those without both)
    $= 220 - 115$ (from Exercises 35 and 37) $= 105$
39. $C(14,6) = 3003$
40. all groups - (those without both types)
    $= 3003 -$ (no bores + no interesting) $= 3003 - (C(8,6) + C(8,6))$
    $= 3003 - 56 = 2947$
41. all - both $= C(14,6) - C(12,4) = 2508$
42. both + neither $= C(12,4) + C(12,6) = 1419$

43. $C(n,r) = \dfrac{n!}{r!(n-r)!} = \dfrac{n!}{(n-r)!(n-(n-r))!} = C(n,n-r)$

44. For any n, k, $C(n,k) = \dfrac{n!}{k!(n-k)!}$
    $C(n-1,k) + C(n-1,k-1) = \dfrac{(n-1)!}{k!(n-1-k)!} + \dfrac{(n-1)!}{(k-1)!(n-1-(k-1))!}$
    $= \dfrac{(n-1)!}{k!(n-1-k)!} + \dfrac{(n-1)!}{(k-1)!(n-k)!} = \dfrac{(n-1)!(n-k)}{k!(n-k)!} + \dfrac{k(n-1)!}{k!(n-k)!}$
    $= \dfrac{(n-1)!(n-k) + k(n-1)!}{k!(n-k)!} = \dfrac{(n-1)!(n-k+k)}{k!(n-k)!} = \dfrac{n!}{k!(n-k)!} = C(n,k)$

45. Proof is by induction on n for a fixed, arbitrary k.
    Basis: n = k: $C(k,k) = C(k+1,k+1)$ true because both equal 1
    Assume $C(k,k) + C(k+1,k) + \ldots + C(n-1,k) = C(n,k+1)$
    Then $C(k,k) + \ldots + C(n-1,k) + C(n,k) = C(n,k+1) + C(n,k)$
    $= C(n,k+1) + [C(n+1,k+1) - C(n,k+1)]$ by Exercise 44
    $= C(n+1,k+1)$

Section 2.4

*1. $(a + b)^4 = C(4,0)a^4 + C(4,1)a^3b + C(4,2)a^2b^2 + C(4,3)ab^3$
    $+ C(4,4)b^4 = a^4 + 4a^3b + 6a^2b^2 + 4ab^3 + b^4$

2. $(x + y)^6 = C(6,0)x^6 + C(6,1)x^5y + C(6,2)x^4y^2 + C(6,3)x^3y^3$
   $+ C(6,4)x^2y^4 + C(6,5)xy^5 + C(6,6)y^6 = x^6 + 6x^5y + 15x^4y^2$
   $+ 20x^3y^3 + 15x^2y^4 + 6xy^5 + y^6$

*3. $(a + 2)^5 = C(5,0)a^5 + C(5,1)a^4(2) + C(5,2)a^3(2)^2$
    $+ C(5,3)a^2(2)^3 + C(5,4)a(2^4) + C(5,5)(2)^5 = a^5 + 10a^4 + 40a^3$
    $+ 80a^2 + 80a + 32$

4. $(a - 4)^4 = C(4,0)a^4 + C(4,1)a^3(-4) + C(4,2)a^2(-4)^2$
   $+ C(4,3)a(-4)^3 + C(4,4)(-4)^4 = a^4 - 16a^3 + 96a^2$
   $- 256a + 256$

*5. $(2x + 3y)^3 = C(3,0)(2x)^3 + C(3,1)(2x)^2(3y) + C(3,2)(2x)(3y)^2$
    $+ C(3,3)(3y)^3 = 8x^3 + 36x^2y + 54xy^2 + 27y^3$

6. $(3x - 1)^5 = C(5,0)(3x)^5 + C(5,1)(3x)^4(-1) + C(5,2)(3x)^3(-1)^2$
   $+ C(5,3)(3x)^2(-1)^3 + C(5,4)(3x)(-1)^4 + C(5,5)(-1)^5$
   $= 243x^5 - 405x^4 + 270x^3 - 90x^2 + 15x - 1$

7. $(2p - 3q)^4 = C(4,0)(2p)^4 + C(4,1)(2p)^3(-3q) + C(4,2)(2p)^2(-3q)^2$
   $+ C(4,3)(2p)(-3q)^3 + C(4,4)(-3q)^4 = 16p^4 - 96p^3q + 216p^2q^2$

$- 216pq^3 + 81q^4$

8. $(3x + \frac{1}{2})^5 = C(5,0)(3x)^5 + C(5,1)(3x)^4(\frac{1}{2}) + C(5,2)(3x)^3(\frac{1}{2})^2$
$+ C(5,3)(3x)^2(\frac{1}{2})^3 + C(5,4)(3x)(\frac{1}{2})^4 + C(5,5)(\frac{1}{2})^5$
$= 243x^5 + \frac{405}{2}x^4 + \frac{132}{2}x^3 + \frac{45}{4}x^2 + \frac{15}{16}x + \frac{1}{32}$

9. $120a^7b^3$
10. $924x^6y^6$
*11. $-489,888x^4$
12. $15,120a^3b^4$
*13. $6561y^8$
14. $729x^6$
*15. $2560x^3y^2$
16. $-1701x^5$
17. $(a + b + c)^3 = ((a + b) + c)^3 = C(3,0)(a + b)^3$
$+ C(3,1)(a + b)^2c + C(3,2)(a + b)c^2 + C(3,3)c^3$
$= [C(3,0)a^3 + C(3,1)a^2b + C(3,2)ab^2 + C(3,3)b^3]$
$+ 3(a^2 + 2ab + b^2)c + 3(a + b)c^2 + c^3 = a^3 + 3a^2b + 3ab^2$
$+ b^3 + 3a^2c + 6abc + 3b^2c + 3ac^2 + 3bc^2 + c^3$
18. $(1 + 0.1)^5 = C(5,0)(1) + C(5,1)(0.1) + C(5,2)(0.1)^2$
$+ C(5,3)(0.1)^3 + C(5,4)(0.1)^4 + C(5,5)(0.1)^5$
$= 1 + 0.5 + 10(0.01) + 10(0.001) + 5(0.0001) + 0.00001$
$= 1 + 0.5 + 0.1 + 0.01 + 0.0005 + 0.00001$
$= 1.61051$
*19. $C(8,1)(2x - y)^7 5^1 = C(8,1)(5)[C(7,4)(2x)^3(-y)^4]$
$= C(8,1)C(7,4)(2)^3(5)x^3y^4 = 11,200x^3y^4$
20. $C(9,2)(x + y)^7(2z)^2 = C(9,2)[C(7,2)x^5y^2](2z)^2$
$= 3024x^5y^2z^2$
21. $C(n,0) - C(n,1) + C(n,2) - \ldots + (-1)^nC(n,n)$
$= (1 + (-1))^n = 0^n = 0$
22. $C(n,0) + C(n,1)2 + C(n,2)2^2 + \ldots + C(n,n)2^n$
$= (1 + 2)^n = 3^n$
23. a) $C(n,0) + C(n,1)x + C(n,2)x^2 + C(n,3)x^3 + \ldots + C(n,n)x^n$
b) Differentiating both sides of the equation
$(1 + x)^n = C(n,0) + C(n,1)x + C(n,2)x^2 + \ldots + C(n,n)x^n$
gives
$n(1 + x)^{n-1} = C(n,1) + 2C(n,2)x + 3C(n,3)x^2 + \ldots + nC(n,n)x^{n-1}$

c) follows from (b) with $x = 1$
d) follows from (b) with $x = -1$

24. $(1 + x)^n = C(n,0) + C(n,1)x + C(n,2)x^2 + C(n,3)x^3 + \ldots + C(n,n)x^n$

Integrating both sides with respect to x,

$$\frac{(1+x)^{n+1}}{n+1} = C(n,0)x + \frac{C(n,1)x^2}{2} + \frac{C(n,2)x^3}{3} + \ldots + \frac{C(n,n)x^{n+1}}{n+1} + C$$

To evaluate constant of integration C, let $x = 0$:

$$\frac{1}{n+1} = 0 + 0 + \ldots + 0 + C$$

so

$$\frac{(1+x)^{n+1}}{n+1} = C(n,0)x + \frac{C(n,1)x^2}{2} + \frac{C(n,2)x^3}{3} + \ldots + \frac{C(n,n)x^{n+1}}{n+1} + \frac{1}{n+1}$$

or

$$\frac{(1+x)^{n+1}-1}{n+1} = C(n,0)x + \frac{C(n,1)x^2}{2} + \frac{C(n,2)x^3}{3} + \ldots + \frac{C(n,n)x^{n+1}}{n+1}$$

Let $x = 1$. Then

$$\frac{2^{n+1}-1}{n+1} = C(n,0) + \frac{1}{2}C(n,1) + \frac{1}{3}C(n,2) + \ldots + \frac{1}{n+1}C(n,n)$$

Let $x = -1$. Then

$$\frac{-1}{n+1} = -C(n,0) + \frac{1}{2}C(n,1) - \frac{1}{3}C(n,2) + \ldots + \frac{(-1)^{n+1}}{n+1}C(n,n)$$

or

$$\frac{1}{n+1} = C(n,0) - \frac{1}{2}C(n,1) + \frac{1}{3}C(n,2) + \ldots + \frac{(-1)^n}{n+1}C(n,n)$$

25. Use induction on n. For $n = 0, 1$, and 2, the result is true. Assume the result is true for $k = n - 1$, and consider $k = n$. By the binomial theorem, the first coefficient is $C(n,0) = 1$; the last coefficient is $C(n,n) = 1$. For $0 < k < n$, the coefficient of $a^{n-k}b^k$ is $C(n,k)$. From Exercise 44, Section 2,3,

$C(n,k) = C(n-1, k-1) + C(n-1, k)$

and $C(n-1, k-1)$ and $C(n-1, k)$ are the two numbers closest to $C(n,k)$ in the previous row, by the inductive hypothesis.

CHAPTER 3

Section 3.1

*1. a) (1,3), (3,3)   b) (4,2), (5,3)
    c) (5,0), (2,2)   d) (1,1), (3,9)

2. a) (1,-1), (-3,3)
   b) 19,41
   c) (3,4,5), (0,5,5), (8,6,10)
   d) (-3,-5), (-4,1/2), (1/2,1/3)

3. *a)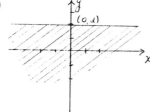

b)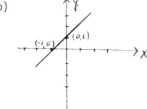

c)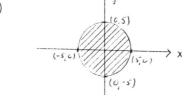

d)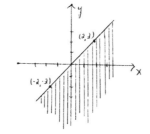

55

4. a) xρy ⟷ x > -1
   b) xρy ⟷ -2 ≤ y ≤ 2
   c) xρy ⟷ x ≤ 2 -y
   d) xρy ⟷ $x^2 + 4y^2 \leq 4$

5. a) (2,6), (3,17), (0,0)   b) (2,12)
   c) none   d) (1,1), (4,8)

6. a) reflexive, antisymmetric
   b) symmetric
   c) symmetric, transitive
   d) reflexive, symmetric, transitive
   e) symmetric, antisymmetric, transitive

7. *a) reflexive, transitive
   *b) reflexive, symmetric, transitive
   *c) symmetric
   d) transitive
   e) reflexive, symmetric, transitive
   f) reflexive, symmetric, transitive
   g) none (not transitive - xρy and yρx ↛ xρx)

8. (b); the equivalence classes are
   [0] = {0,3,6,9,⋯,-3,-6,-9,⋯}
   [1] = {1,4,7,10,⋯,-2,-5,-8,⋯}
   [2] = {2,5,8,11,⋯,-1,-4,-7,⋯}

   (e); the equivalence classes are sets consisting of squares with equal length sides

   (f); the equivalence classes are sets consisting of strings with the same number of characters

9. For example:
   a) S = set of all lines in the plane, xρy ⟷ x coincides with y or x is perpendicular to y. Then ρ is reflexive (x coincides with x) and symmetric (x coincides with y ⟶ y coincides with x or x ⊥ y ⟶ y ⊥ x) but not transitive (x ⊥ y and y ⊥ z only implies x and z parallel)
   b) S = set of integers, xρy ⟷ $x^2 \leq y^2$. Then ρ is reflexive ($x^2 \leq x^2$) and transitive ($x^2 \leq y^2$ and $y^2 \leq z^2 \longrightarrow x^2 \leq z^2$) but not symmetric (2ρ3 but not 3ρ2).
   c) S = set of nonnegative integers, xρy ⟷ x < y. Then ρ is

not reflexive (x is not less than x), not symmetric
(x < y $\not\to$ y < x), but is transitive (x < y and y < z $\to$
x < z).

d) S = set of integers, xρy $\leftrightarrow$ x $\leq$ |y|. Then ρ is reflexive
(x $\leq$ |x|), but not symmetric (-2ρ3 but not 3ρ-2) and not
transitive (3ρ-4 and -4ρ2 but not 3ρ2).

10. a) yes, yes   b) yes, yes
    c) no, yes   d) no, yes

11. *a) ρ = {(1,1)}
    b) ρ = {(1,2), (2,1), (1,3)}
    c) Assume ρ is asymmetric and not irreflexive. Then for
       some xϵS, (x,x)$\epsilon\rho$. But then (x,x)$\epsilon\rho$ and (x,x)$\epsilon\rho$, which
       contradicts asymmetry.
    d) Assume ρ is irreflexive and transitive but not asymmetric.
       Then for some x,yϵS, (x,y)$\epsilon\rho$ and (y,x)$\epsilon\rho$. By transitivity,
       (x,x)$\epsilon\rho$, which contradicts irreflexivity.
    e) Assume ρ is nonempty, symmetric, and transitive. Let
       (x,y)$\epsilon\rho$. Then (y,x)$\epsilon\rho$ by symmetry, and (x,x)$\epsilon\rho$ by
       transitivity. Therefore (x,x)$\epsilon\rho$ for some xϵS and ρ is
       not irreflexive.

12. a) Let xϵ#A. Then xρy for all yϵA, so by symmetry, yρx for
       all yϵA, and xϵA#. Therefore #A $\subseteq$ A#. By a similar
       argument, A# $\subseteq$ #A and #A = A#.
    b) Assume A $\subseteq$ B and let xϵ#B. Then xρy for all yϵB and
       since A $\subseteq$ B, xρy for all yϵA. Thus xϵ#A, and #B $\subseteq$ #A.
       Similarly B# $\subseteq$ A#.
    c) Let xϵA and let zϵ#A. Then zρy for all yϵA, so in
       particular, zρx. Since z was arbitrary, zρx holds for
       all z in #A, and therefore xϵ(#A)# and A $\subseteq$ (#A)#.
    d) Let xϵA and let zϵA#. Then yρz for all yϵA, so in
       particular, xρz. Since z was arbitrary, xρz holds for
       all z in A#, and therefore xϵ#(A#) and A $\subseteq$ #(A#).

13. a)
```
    • c
    |
    • b
    |
    • a
```

b)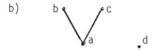

c)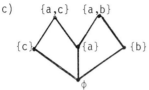

14. a) a is minimal and least
       c is maximal and greatest
    b) a and d are minimal
       b, c, and d are maximal
    c) φ is minimal and least
       {a,c} and {a,b} are maximal

15. 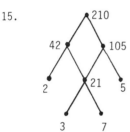 No least element; minimal elements of 2, 3, 5, 7; greatest element = maximal element = 210. Totally ordered subsets: {3,21,105,210}, {3,21,42,210}, {7,21,105,210}, {7,21,42,210}.

*16. a) 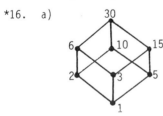   b)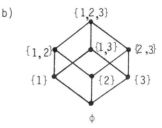

The two graphs are identical in structure.

17. *a) ρ = {(1,1), (2,2), (3,3), (4,4), (5,5), (1,3), (3,5), (1,5), (2,4), (4,5), (2,5)}
    b) ρ = {(a,a), (b,b), (c,c), (d,d), (e,e), (f,f), (a,d), (b,e), (c,f)}
    c) ρ = {(1,1), (2,2), (3,3), (4,4), (5,5), (1,2), (2,4), (4,5), (1,4), (1,5), (2,5), (1,3), (3,4), (3,5)}

18. Reflexive: $(s_1,t_1)\mu(s_1,t_1)$ because both $s_1\rho s_1$ and $t_1\sigma t_1$ due to reflexivity of ρ and σ.
Antisymmetric: $(s_1,t_1)\mu(s_2,t_2)$ and $(s_2,t_2)\mu(s_1,t_1) \rightarrow s_1\rho s_2$ and $s_2\rho s_1$, $t_1\sigma t_2$ and $t_2\sigma t_1 \rightarrow s_1 = s_2$ and $t_1 = t_2$ due to antisymmetry of ρ and $\sigma \rightarrow (s_1,t_1) = (s_2,t_2)$.
Transitive: $(s_1,t_1)\mu(s_2,t_2)$ and $(s_2,t_2)\mu(s_3,t_3) \rightarrow s_1\rho s_2$ and $s_2\rho s_3$, $t_1\sigma t_2$ and $t_2\sigma t_3 \rightarrow s_1\rho s_3$ and $t_1\sigma t_3$ due to transitivity of ρ and $\sigma \rightarrow (s_1,t_1)\mu(s_3,t_3)$.

19. a) Reflexive: For $x \in S$, $x \preceq x$ because $\preceq$ is reflexive, and $x \preceq x \rightarrow x \succeq x$.
Antisymmetric: $x \succeq y$ and $y \succeq x \rightarrow y \preceq x$ and $x \preceq y \rightarrow x = y$ (by antisymmetry of $\preceq$).
Transitive: $x \succeq y$ and $y \succeq z \rightarrow y \preceq x$ and $z \preceq y \rightarrow z \preceq x$ (by transitivity of $\preceq$) $\rightarrow x \succeq z$.

b)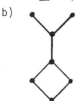

c) Let $(x,y) \in (\succeq - X)$. Then $x \succeq y$ and $x \neq y$. Therefore $y \preceq x$ and $(y,x) \in \preceq$. If $(x,y) \in \preceq$, then by antisymmetry of $\preceq$, $x = y$, a contradiction. Thus $(x,y) \in \preceq'$. This proves that $(\succeq - X) \subseteq \preceq'$. Now let $(x,y) \in \preceq'$. Then not $x \preceq y$, so by total ordering, $y \preceq x$, and $x \succeq y$. If $x = y$ then $x \preceq y$ by reflexivity, a contradiction. Thus $(x,y) \in (\succeq - X)$, and $\preceq' \subseteq (\succeq - X)$.

20. a) Reflexive: $X \preceq X$ by (2) with $m = k$
Antisymmetric: Let $X \preceq Y$ and $Y \preceq X$. If $X \neq Y$, let $m + 1$ be the first index where $x_{m+1} \neq y_{m+1}$. Then $x_{m+1} \preceq y_{m+1}$ and $y_{m+1} \preceq x_{m+1} \Rightarrow x_{m+1} = y_{m+1}$, a contradiction.
Transitive: Let $X \preceq Y$ and $Y \preceq Z$. Then $x_p \preceq y_p$ for some $p \leq k$ and $y_q \preceq z_q$ for some $q \leq k$. Let $m = \min(p,q)$. Then $x_m \preceq z_m$ and $X \preceq Z$. Total: by "otherwise"
b) bah $\prec$ be $\prec$ boo $\prec$ bug $\prec$ bugg

*21. a) when; no; all but the last
b)

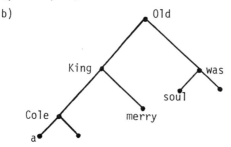

Maximal elements: a, merry, soul

22. The enumeration of A* is a, b, ..., z, aa, ab, ac, ..., az, ba, bb, ..., bz, ca, ..., cz, ..., za, ..., zz, aaa, aab, aac, ... (each interval is finite)

*23. a) [a] = {a,c} = [c]
b) [3] = {1,2,3}
   [4] = {4,5}
c) [1] = {... -5,-3,-1,1,3,5,...}
d) [-3] = {...,-13,-8,-3,2,7,12,...}

24. a) {(1,1), (2,2), (3,3), (4,4), (1,2), (2,1), (3,4), (4,3)}
b) {(a,a), (b,b), (c,c), (d,d), (e,e), (a,b), (b,a), (a,c), (c,a), (b,c), (c,b), (d,e), (e,d)}

25. a) all in one block + each element separate + 2 in one block, 1 in other = 1 + 1 + C(3,2) = 1 + 1 + 3 = 5
b) all in one block + each element separate + 1 and 3 + 2 and 2 + 1 and 1 and 2 = 1 + 1 + C(4,3) + C(4,2)/2 + C(4,2) = 1 + 1 + 4 + 3 + 6 = 15

26. Reflexive: $\pi_1 \leq \pi_1$ because each block of π_1 is a subset of itself due to reflexivity of set inclusion. Antisymmetry and transitivity also follow from the corresponding set inclusion properties.

27. Clearly P ⟷ P is a tautology. If P ⟷ Q is a tautology, then P and Q have the same truth values everywhere, so Q ⟷ P is a tautology. If P ⟷ Q and Q ⟷ R are tautologies, then P,Q, and R have the same truth values everywhere, and P ⟷ R is a tautology. The equivalence classes are sets consisting of statements with the same truth values everywhere.

28. a) The EQUIVALENCE statement binds one or more variables to the same location in the address space. Thus for any variable names X,Y, and Z, $X \sim X$, $X \sim Y \longrightarrow Y \sim X$, and $X \sim Y$, $Y \sim Z \longrightarrow X \sim Z$. The equivalence classes are sets consisting of variable names assigned to the same memory location.
    b) Same as part (a).

Section 3.2

1. a) not a function from S to T (not a subset of S x T)
   b) function
   c) function; one-to-one and onto
   d) not a function from S to T (0 has no associated value)
   e) not a function (two values associated with 0)
2. For part (c), $f^{-1}: T \rightarrow S$, $f^{-1} = \{(3,2), (7,4), (1,0), (5,6)\}$
*3. a) function
   b) not a function; undefined at x = 0
   c) function; onto
   d) bijection; $f^{-1}: \{p,q,r\} \longrightarrow \{1,2,3\}$ where
       $f^{-1} = \{(q,1), (r,2), (p,3)\}$
   e) function; one-to-one
   f) bijection; $h^{-1}: R^2 \longrightarrow R^2$ where
       $h^{-1}(x,y) = (y-1, x-1)$
4. a) The program reads all 16 data records; it reads the fields from 1 to K, and prints K. Here K = I, and I runs from 1 to 16. Output is 1,2,$\cdots$,16. The function is thus (record1,1), (record2,2),$\cdots$, (record16,16). It is a bijection.
   b) Here K = I, I runs from 1 to 15. The function is (record1,1), $\cdots$, (record15,15); it is injective but not surjective, as 16 is not in the range.
   c) The function here is (record1,2), (record2,2), (record3,3), (record4,3),$\cdots$, (record16,9); it is neither surjective nor injective.
5. For example:
   a) $f = \{(a,x), (b,x), (c,y), (d,y)\}$
   b) $f = \{(a,x), (b,x), (c,y), (d,z)\}$
   c) no

6. a) Let $t \in f(A \cap B)$. Then $t = f(s)$ for some $s \in A \cap B$. Thus $t \in f(A)$ and $t \in f(B)$, so $t \in f(A) \cap f(B)$.

   b) Assume f is one-to-one. By (a), $f(A \cap B) \subseteq f(A) \cap f(B)$. Let $t \in f(A) \cap f(B)$. Then $t = f(s_1)$ for some $s_1 \in A$ and $t = f(s_2)$ for some $s_2 \in B$. Because f is one-to-one, $s_1 = s_2$ and $s_1 \in A \cap B$, so $t \in f(A \cap B)$. Thus $f(A) \cap f(B) \subseteq f(A \cap B)$ and the two sets are equal.

   Now assume $f(A \cap B) = f(A) \cap f(B)$ for all subsets A and B of S, and let $s_1, s_2 \in S$ such that $f(s_1) = f(s_2)$. Let $t = f(s_1)$ and let $A = \{s_1\}$, $B = \{s_2\}$. Then $t \in f(A) \cap f(B)$, or $t \in f(A \cap B)$. Therefore $A \cap B \neq \phi$, and $s_1 = s_2$, so f is one-to-one.

*7. a) $2^3 = 8$; 6

   b) n^m

   c) $n(n-1)(n-2)\cdots(n-(m-1)) = \dfrac{n!}{(n-m)!}$

8. a) Assume f is onto. If two distinct elements of S map to one element of S, then n - 2 elements are left to map onto n - 1 elements, which cannot be done. Therefore f is one-to-one. Now assume f is one-to-one. Then the n elements of S map to n distinct elements of S; thus every element of S is in the range of f, and f is onto.

   b) For example, $S = N$, $f: N \rightarrow N$ given by $f(x) = 2x$.

   c) For example, $S = N$, $f: N \rightarrow N$ given by $\begin{cases} f(0) = 0 \\ f(x) = x-1, x \geq 1 \end{cases}$

*9. $\{m,n,o,p\}$; $\{n,o,p\}$; $\{o\}$

10. $g \circ f = \{(1,6), (2,7), (3,9), (4,9)\}$

*11. a) $(g \circ f)(5) = g(f(5)) = g(6) = 18$

   b) $(f \circ g)(5) = f(g(5)) = f(15) = 16$

   c) $(g \circ f)(x) = g(f(x)) = g(x+1) = 3(x+1) = 3x+3$

   d) $(f \circ g)(x) = f(g(x)) = f(3x) = 3x+1$

   e) $(f \circ f)(x) = f(f(x)) = f(x+1) = (x+1)+1 = x+2$

   f) $(g \circ g)(x) = g(g(x)) = g(3x) = 3(3x) = 9x$

12. a) 25   b) 1210 (the index doesn't count)   c) 9

13. a) (1, 3, 5, 2)

    b) (1, 4, 3, 2, 5)

14. a) $\begin{pmatrix} a & b & c & d \\ c & b & d & a \end{pmatrix}$
    b) $\begin{pmatrix} a & b & c & d \\ b & d & a & c \end{pmatrix}$
    c) $\begin{pmatrix} a & b & c & d \\ d & a & c & b \end{pmatrix}$
    d) $\begin{pmatrix} a & b & c & d \\ c & d & b & a \end{pmatrix}$

*15. a) $\begin{pmatrix} 1 & 2 & 3 \\ 1 & 2 & 3 \end{pmatrix}$ $\begin{pmatrix} 1 & 2 & 3 \\ 1 & 3 & 2 \end{pmatrix}$ $\begin{pmatrix} 1 & 2 & 3 \\ 3 & 2 & 1 \end{pmatrix}$
    $\begin{pmatrix} 1 & 2 & 3 \\ 3 & 1 & 2 \end{pmatrix}$ $\begin{pmatrix} 1 & 2 & 3 \\ 2 & 1 & 3 \end{pmatrix}$ $\begin{pmatrix} 1 & 2 & 3 \\ 2 & 3 & 1 \end{pmatrix}$
    b) n!

16. Both h ∘ (g ∘ f) and (h ∘ g) ∘ f have domain and codomain A. For x ε A, (h ∘ (g ∘ f))(x) = h((g ∘ f)(x)) = h(g(f(x))) = (h ∘ g)(f(x)) = ((h ∘ g) ∘ f)(x).

*17. a) (1,2,5,3,4)   b) (1,7,8) ∘ (2,4,6)
    c) (1,5,2,4) ∘ (3,6)   d) (2,3) ∘ (4,8) ∘ (5,7)

18. a) $\begin{pmatrix} 1 & 2 & 3 & 4 \\ 2 & 1 & 4 & 3 \end{pmatrix}$ b) For example, $\begin{pmatrix} 1 & 2 & 3 & 4 \\ 3 & 1 & 2 & 4 \end{pmatrix}$

*19. a) $f^{-1}(x) = \frac{x}{2}$
    b) $f^{-1}(x) = \sqrt[3]{x}$
    c) $f^{-1}(x) = 3x - 4$

20. a) Assume f has a left inverse g, and that $f(s_1) = f(s_2)$. Then $g(f(s_1)) = g(f(s_2))$ or $(g \circ f)(s_1) = (g \circ f)(s_2)$ and $i_S(s_1) = i_S(s_2)$, thus $s_1 = s_2$ and f is one-to-one. Now let f: S→T with f one-to-one. We want to define g: T→S. For t ε T with t ε f(S), define g(t) to be the unique preimage of t under f. For t ε T with t ∉ f(S), let g(t) be any fixed element of S. Then g: T→S and for s ε S, (g ∘ f)(s) = g(f(s)) = g(t) = s, so g ∘ f = i_S.

    b) Assume f has a right inverse g, and let t ε T. Then t = $i_T(t)$ = (f ∘ g)(t) = f(g(t)); g(t) ε S, so t ε f(S) and f is onto. Now let f: S→T with f onto. Then every t ε T has at least one preimage in S under f. Define g: T→S by g(t) = a fixed preimage s of t. Then

$(f \circ g)(t) = f(g(t)) = f(s) = t$, so $f \circ g = i_T$.

c) For example:
$$g_1(x) = \begin{cases} x/3 & \text{for } x = 3k, \quad k \text{ an integer} \\ 0 & \text{for } x \neq 3k \end{cases}$$
$$g_2(x) = \begin{cases} x/3 & \text{for } x = 3k, \quad k \text{ an integer} \\ 1 & \text{for } x \neq 3k \end{cases}$$

21. a) If $f(s_1) = f(s_2)$ then $g(f(s_1)) = g(f(s_2))$ so $(g \circ f)(s_1) = (g \circ f)(s_2)$. Because $g \circ f$ is one-to-one, $s_1 = s_2$ and therefore f is one-to-one.

    b) For $u \in U$, there exists $s \in S$ such that $(g \circ f)(s) = u$, because $g \circ f$ is onto. Thus $g(f(s)) = u$ and $f(s)$ is a member of T that is a preimage of u under g, and g is onto.

    c) Let $S = \{1,2,3\}$, $T = \{1,2,3,4\}$, $U = \{1,2,3\}$, $f = \{(1,1), (2,2), (3,3)\}$, $g = \{(1,1), (2,2), (3,3), (4,3)\}$. Then $f: S \to T$, $g: T \to U$, g is not one-to-one but $g \circ f = \{(1,1), (2,2), (3,3)\}$ is one-to-one.

    d) same example as for (c)

22. $f^{-1}: T \to S$, $g^{-1}: U \to T$, so $f^{-1} \circ g^{-1}: U \to S$. For $s \in S$, let $f(s) = t$ and $g(t) = u$. Then $(f^{-1} \circ g^{-1}) \circ (g \circ f)(s) = f^{-1}(g^{-1}(u)) = f^{-1}(t) = s$. Also for $u \in U$, $(g \circ f) \circ (f^{-1} \circ g^{-1})(u) = g(f(s)) = g(t) = u$. Then $(f^{-1} \circ g^{-1}) \circ (g \circ f) = i_S$ and $(g \circ f) \circ (f^{-1} \circ g^{-1}) = i_U$, so $f^{-1} \circ g^{-1} = (g \circ f)^{-1}$.

23. Reflexive: $S \rho S$ by the identity function.
    Symmetric: If $S \rho T$ and f is a bijection from S to T, then $f^{-1}: T \to S$ and f^{-1} is a bijection, so $T \rho S$.
    Transitive: If $S \rho T$ and $T \rho U$, $f: S \to T$, $g: T \to U$, f and g bijections, then $g \circ f: S \to U$ and $g \circ f$ is a bijection, so $S \rho U$.

24. a) For $x \in S$, $f(x) = f(x)$, so $x \rho x$ and ρ is reflexive.
    For $x, y \in S$, if $x \rho y$ then $f(x) = f(y)$ and $f(y) = f(x)$ so $y \rho x$ and ρ is symmetric.
    For $x, y, z \in S$, if $x \rho y$ and $y \rho z$ then $f(x) = f(y)$ and $f(x) = f(z)$ so $f(x) = f(z)$, and $x \rho z$, so ρ is transitive.

    b) $[4] = \{4, -4\}$

Section 3.3

*1. 2, -4
2. x = 2, y = 4
*3. x = 1, y = 3, z = -2, w = 4
4. u = 1, v = -3, w = 7
*5. $\begin{bmatrix} 6 & -5 \\ 0 & 3 \\ 5 & 3 \end{bmatrix}$
6. $\begin{bmatrix} -2 & 7 \\ -2 & -3 \\ 1 & 5 \end{bmatrix}$
7. $\begin{bmatrix} 12 & 3 & 6 \\ 18 & -3 & 15 \\ 3 & 9 & 6 \end{bmatrix}$
8. $\begin{bmatrix} -4 & -8 \\ -12 & 2 \end{bmatrix}$
*9. $\begin{bmatrix} 14 & -17 \\ 2 & 9 \\ 9 & 1 \end{bmatrix}$
10. not possible
11. $\begin{bmatrix} 18 & -15 \\ 0 & 9 \\ 15 & 9 \end{bmatrix}$
12. $\begin{bmatrix} -12 & -24 \\ -36 & 6 \end{bmatrix}$
*13. $\begin{bmatrix} 21 & -23 \\ 33 & -44 \\ 11 & 1 \end{bmatrix}$
14. $\begin{bmatrix} -28 & 22 \\ 20 & 1 \\ -2 & 9 \end{bmatrix}$
15. $\begin{bmatrix} 10 & 7 \\ -2 & -4 \\ 30 & 8 \end{bmatrix}$
16. not possible

*17. $\begin{bmatrix} 28 & 4 \\ 6 & 25 \end{bmatrix}$

18. $\begin{bmatrix} 17 & 6 \\ 29 & 29 \\ 7 & 8 \end{bmatrix}$

19. $A \cdot B = \begin{bmatrix} 10 & 4 \\ 18 & -3 \end{bmatrix} \quad B \cdot A = \begin{bmatrix} 14 & 1 \\ 4 & -7 \end{bmatrix}$

20. $A(B \cdot C) = \begin{bmatrix} 3 & -1 \\ 2 & 5 \end{bmatrix} \begin{bmatrix} 26 & -22 \\ 10 & -8 \end{bmatrix} = \begin{bmatrix} 64 & -58 \\ 102 & -84 \end{bmatrix}$

$(A \cdot B)C = \begin{bmatrix} 10 & 4 \\ 18 & -3 \end{bmatrix} \begin{bmatrix} 6 & -5 \\ 2 & -2 \end{bmatrix} = \begin{bmatrix} 68 & -58 \\ 102 & -84 \end{bmatrix}$

*21. $A(B + C) = \begin{bmatrix} 3 & -1 \\ 2 & 5 \end{bmatrix} \begin{bmatrix} 10 & -4 \\ 4 & -3 \end{bmatrix} = \begin{bmatrix} 26 & -9 \\ 40 & -23 \end{bmatrix}$

$A \cdot B + A \cdot C = \begin{bmatrix} 10 & 4 \\ 18 & -3 \end{bmatrix} + \begin{bmatrix} 16 & -13 \\ 22 & -20 \end{bmatrix} = \begin{bmatrix} 26 & -9 \\ 40 & -23 \end{bmatrix}$

22. $(A + B)C = \begin{bmatrix} 7 & 0 \\ 4 & 4 \end{bmatrix} \begin{bmatrix} 6 & -5 \\ 2 & -2 \end{bmatrix} = \begin{bmatrix} 42 & -35 \\ 32 & -28 \end{bmatrix}$

$A \cdot C + B \cdot C = \begin{bmatrix} 16 & -13 \\ 22 & -20 \end{bmatrix} + \begin{bmatrix} 26 & -22 \\ 10 & -8 \end{bmatrix} = \begin{bmatrix} 42 & -35 \\ 32 & -28 \end{bmatrix}$

23. $x = 3, y = 4$

24. $I \cdot A = A$ for any n x n matrix A, in particular, if $A = I$, then $I \cdot I = I$

25.*a) $\begin{bmatrix} 1 & 3 \\ 2 & 2 \end{bmatrix} \begin{bmatrix} -\frac{1}{2} & \frac{3}{4} \\ \frac{1}{2} & -\frac{1}{4} \end{bmatrix} = \begin{bmatrix} 1 & 0 \\ 0 & 1 \end{bmatrix} = \begin{bmatrix} -\frac{1}{2} & \frac{3}{4} \\ \frac{1}{2} & -\frac{1}{4} \end{bmatrix} \begin{bmatrix} 1 & 3 \\ 2 & 2 \end{bmatrix}$

*b) For $\begin{bmatrix} 1 & 2 \\ 2 & 4 \end{bmatrix} \begin{bmatrix} b_{11} & b_{12} \\ b_{21} & b_{22} \end{bmatrix} = \begin{bmatrix} 1 & 0 \\ 0 & 1 \end{bmatrix}$

$b_{11} + 2b_{21} = 1 \quad b_{12} + 2b_{22} = 0$

$2b_{11} + 4b_{21} = 0 \quad 2b_{12} + 4b_{22} = 1$

which is an inconsistent system of equations with no solution.

c) $\begin{bmatrix} a_{11} & a_{12} \\ a_{21} & a_{22} \end{bmatrix} \begin{bmatrix} b_{11} & b_{12} \\ b_{21} & b_{22} \end{bmatrix} = \begin{bmatrix} 1 & 0 \\ 0 & 1 \end{bmatrix}$

implies

$a_{11}b_{11} + a_{12}b_{21} = 1 \qquad a_{11}b_{12} + a_{12}b_{22} = 0$

$a_{21}b_{11} + a_{22}b_{21} = 0 \qquad a_{21}b_{12} + a_{22}b_{22} = 1$

Solving these systems of equations gives

$$b_{11} = \frac{a_{22}}{a_{11}a_{22} - a_{12}a_{21}} \qquad b_{12} = \frac{-a_{12}}{a_{11}a_{22} - a_{12}a_{21}}$$

$$b_{21} = \frac{-a_{21}}{a_{11}a_{22} - a_{12}a_{21}} \qquad b_{22} = \frac{a_{11}}{a_{11}a_{22} - a_{12}a_{21}}$$

These values can all be found if $a_{11}a_{22} - a_{12}a_{21} \neq 0$.

26. a) $A^T = \begin{bmatrix} 1 & 6 \\ 3 & -2 \\ 4 & 1 \end{bmatrix}$

b) If A is symmetric then $a_{ij} = a_{ji}$ and $A^T(i,j) = A(j,i) = A(i,j)$. Therefore $A^T = A$. If $A^T = A$, then $A(i,j) = A^T(i,j) = A(j,i)$ and A is symmetric.

c) Let $A + B = C$. Then $C^T(i,j) = C(j,i) = A(j,i) + B(j,i)$
$= A^T(i,j) + B^T(i,j)$ and $C^T = A^T + B^T$

d) Let A be an n x m matrix and B be an m x p matrix; then A^T is m x n and B^T is p x m. Let $A \cdot B = C$.
Then $C^T(i,j) = C(j,i) = \sum_{k=1}^{m} a_{jk}b_{ki} = \sum_{k=1}^{m} A^T(k,j)B^T(i,k)$
$= \sum_{k=1}^{m} B^T(i,k)A^T(k,j) = (B^T \cdot A^T)(i,j)$ and $C^T = B^T \cdot A^T$

*27. For example,

$$\begin{bmatrix} 1 & 1 \\ -1 & -1 \end{bmatrix} \begin{bmatrix} 1 & 1 \\ -1 & -1 \end{bmatrix} = \begin{bmatrix} 0 & 0 \\ 0 & 0 \end{bmatrix}$$

28. For example,

$$\begin{bmatrix} 1 & 2 \\ 2 & 1 \end{bmatrix} \begin{bmatrix} 1 & 2 \\ 1 & 2 \end{bmatrix} = \begin{bmatrix} 2 & 1 \\ 1 & 2 \end{bmatrix} \begin{bmatrix} 1 & 2 \\ 1 & 2 \end{bmatrix} = \begin{bmatrix} 3 & 6 \\ 3 & 6 \end{bmatrix}$$

but

$$\begin{bmatrix} 1 & 2 \\ 2 & 1 \end{bmatrix} \neq \begin{bmatrix} 2 & 1 \\ 1 & 2 \end{bmatrix}$$

CHAPTER 4

Section 4.1

*1. (d) - the two nodes of degree 3 are not adjacent.

*2. a) This is a tree   b)   c) For example,

*3. a)   b)   c)

d) $\dfrac{n(n-1)}{2}$

e) The number of arcs is $C(n,2) = \dfrac{n(n-1)}{2}$. (Other proof methods include induction on the number of nodes.)

4. a) yes   b) no   c) b   d) does not exist   e) 2   f) 3

*5. 4, 5, 6; length 2; for example (naming the nodes),
1-2-1-2-2-1-4-5-6

6.

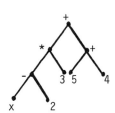

*7.

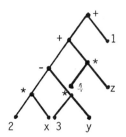

8.

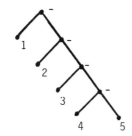

9.

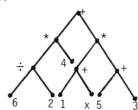

10. Proof is by induction on d. For d = 0, the only node is the root, and $2^0 = 1$. Assume that there are at most 2^d nodes at depth d, and consider depth d + 1. There are at most two children for each node at depth d, so the maximum number of nodes at depth d + 1 is $2 \cdot 2^d = 2^{d+1}$.

11. a)

7 nodes

b)

15 nodes

c) $2^{h+1} - 1$

d) Proof is by induction on h. For h = 0, the tree consists only of the root, so the number of nodes $= 1 = 2^{0+1} - 1$.

Assume that a full binary tree of height h has $2^{h+1} - 1$ nodes. Consider a full binary tree of height h + 1. The leaves are at depth h + 1, and there are 2^h nodes at depth h (by Exercise 10). Removing the leaves and associated arcs gives a full binary tree of height h, with $2^{h+1} - 1$ nodes, by the inductive assumption. There are 2 leaves in the original tree for each of the 2^h leaves in the reduced tree, so the total number of nodes in the original tree is
$2^{h+1} - 1 + 2(2^h) = 2^{h+1} - 1 + 2^{h+1} = 2 \cdot 2^{h+1} - 1 = 2^{h+2} - 1$.

12.

*13.

14.

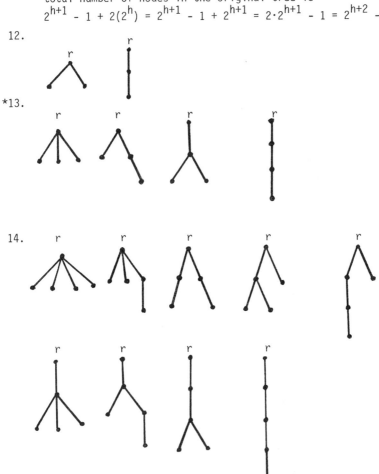

*15. Assume that the graph is a tree; then it is acyclic and connected. For any two nodes x and y, a path exists from x to y because the graph is connected. If the path is not unique, then the two paths diverge at some node n_1 and converge at some node n_2, and there is a cycle from n_1 through n_2 and back to n_1, which contradicts the fact that the graph is acyclic. Now suppose there is a unique path between any two nodes. Then the graph is connected, and there are no cycles because the presence of a cycle produces a non-unique path between two nodes on the cycle. Hence the graph is a tree.

16. In a full binary tree, all internal nodes have two children, so the total number of "children nodes" is 2x; the only "non-child" node is the root, so there is a total of 2x + 1 nodes. Because x of these are internal, x + 1 are leaves.

17. Let x be the number of nodes with two children; we want to show there are x + 1 leaves; one proof is by induction on x. If x = 0, then the tree must be a "chain, i.e., it looks like and there is only 1 leaf; 1 = x + 1. Assume in any binary tree with x nodes having two children that there are x + 1 leaves. Now consider a binary tree where x + 1 nodes have two children. Remove the subtree rooted by one of the children of a node with two children at the maximum depth at which such a node occurs, for example:

This subtree has no nodes with two children, and exactly one leaf. The remaining tree has x nodes with two children and, by the inductive hypothesis, x + 1 leaves. Thus the original tree had x + 2 leaves.

Another proof parallels that of Exercise 16. Let x = the number of nodes with two children, let y = the number of nodes with one child. Then the total number of nodes is 2x + y + 1, and the number of internal nodes is x + y, so the number of leaves is 2x + y + 1 - (x + y) = x + 1.

18. If G is a tree, then G is connected. Suppose we remove an arc a between n_1 and n_2 and G remains connected. Then there is a path from n_1 to n_2. Adding a to this path results in a cycle from n_1 to n_1, which contradicts the definition of a tree.

Suppose G is connected and removing any single arc makes G unconnected. Then there is a unique path between any two nodes and the graph is a tree (Exercise 15).

19. Let G be a tree and add an arc a between n_1 and n_2. Because G was originally connected, there was a path between n_1 and n_2; adding a to this path results in a cycle from n_1 to n_1.
If adding one arc to G results in a graph with exactly one cycle, then the original graph was acyclic and connected, a tree.

20. $n = 4$, $a = 5$, $r = 2$ and $n - a + r = 4 - 5 + 2 = 1$.

21. For $a = 0$, the graph consists of a single node. Then $a = 0$, $n = 1$, $r = 0$, and $n - a + r = 1$. Now assume that in any connected planar graph with k arcs, $n - k + r = 1$, and consider a graph with $k + 1$ arcs. A graph with $k + 1$ arcs can be obtained from a graph with k arcs in one of two ways: (1) add a new node and connect it by a new arc to an old node, or (2) connect two old nodes by a new arc. In case (1), there is one new node and one new arc, and the number of regions stays the same. Therefore $n - a + r$ stays the same. In case (2), there is one new arc and one new region added, so $n - a + r$ stays the same. In both cases $n - a + r$ remains 1.

Section 4.2

*1. $\begin{bmatrix} 1 & 1 & 0 & 0 & 2 \\ 1 & 1 & 1 & 1 & 1 \\ 0 & 1 & 0 & 1 & 0 \\ 0 & 1 & 1 & 0 & 0 \\ 2 & 1 & 0 & 0 & 0 \end{bmatrix}$

2. $\begin{bmatrix} 1 & 0 & 1 & 0 & 0 \\ 0 & 0 & 1 & 1 & 1 \\ 1 & 1 & 0 & 1 & 0 \\ 0 & 1 & 1 & 0 & 1 \\ 0 & 1 & 0 & 1 & 0 \end{bmatrix}$

3. $\begin{bmatrix} 0 & 1 & 1 & 0 & 0 & 0 & 0 \\ 1 & 0 & 0 & 1 & 0 & 1 & 1 \\ 1 & 0 & 0 & 0 & 1 & 1 & 1 \\ 0 & 1 & 0 & 0 & 0 & 1 & 0 \\ 0 & 0 & 1 & 0 & 0 & 0 & 1 \\ 0 & 1 & 1 & 1 & 0 & 0 & 1 \\ 0 & 1 & 1 & 0 & 1 & 1 & 0 \end{bmatrix}$

4. $\begin{bmatrix} 0 & 1 & 1 & 0 & 0 & 0 \\ 1 & 0 & 1 & 0 & 0 & 0 \\ 1 & 1 & 0 & 0 & 0 & 0 \\ 0 & 0 & 0 & 0 & 1 & 1 \\ 0 & 0 & 0 & 1 & 0 & 1 \\ 0 & 0 & 0 & 1 & 1 & 0 \end{bmatrix}$

*5. $\begin{bmatrix} 0 & 1 & 0 & 0 \\ 0 & 0 & 1 & 1 \\ 0 & 0 & 0 & 1 \\ 0 & 0 & 1 & 0 \end{bmatrix}$

6. $\begin{bmatrix} 0 & 1 & 1 & 0 \\ 1 & 1 & 1 & 1 \\ 0 & 0 & 0 & 1 \\ 0 & 0 & 0 & 0 \end{bmatrix}$

7. $A^2 = \begin{bmatrix} 2 & 0 & 0 & 1 & 1 & 2 & 2 \\ 0 & 4 & 3 & 1 & 1 & 2 & 1 \\ 0 & 3 & 4 & 1 & 1 & 1 & 2 \\ 1 & 1 & 1 & 2 & 0 & 1 & 2 \\ 1 & 1 & 1 & 0 & 2 & 2 & 1 \\ 2 & 2 & 1 & 1 & 2 & 4 & 2 \\ 2 & 1 & 2 & 2 & 1 & 2 & 4 \end{bmatrix}$

8. 13; $A^3 = \begin{bmatrix} 16 & 15 & 5 & 5 & 16 \\ 15 & 16 & 7 & 7 & 13 \\ 5 & 7 & 3 & 4 & 4 \\ 5 & 7 & 4 & 3 & 4 \\ 16 & 13 & 4 & 4 & 9 \end{bmatrix}$

*9. $R = \begin{bmatrix} 2 & 4 & 4 & 4 \\ 4 & 6 & 6 & 7 \\ 0 & 0 & 0 & 1 \\ 0 & 0 & 0 & 0 \end{bmatrix}$

10. $R = \begin{bmatrix} 7 & 12 & 13 & 12 & 13 & 12 \\ 12 & 19 & 20 & 17 & 19 & 16 \\ 0 & 0 & 2 & 3 & 5 & 3 \\ 0 & 0 & 1 & 1 & 3 & 2 \\ 0 & 0 & 2 & 1 & 1 & 3 \\ 0 & 0 & 3 & 2 & 3 & 2 \end{bmatrix}$

11.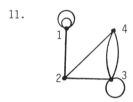

12.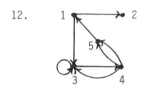

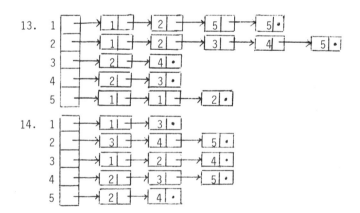

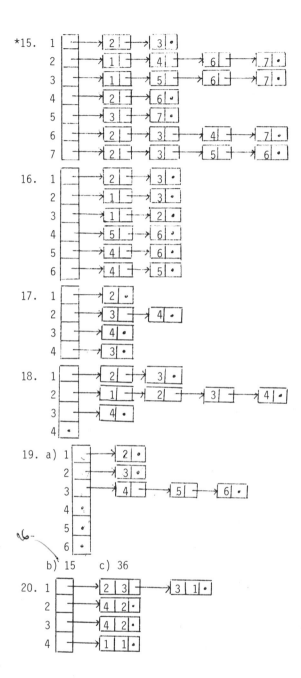

* 21.

Node		Pointer
1		5
2		7
3		11
4		0
5	2	6
6	3	0
7	1	8
8	2	9
9	3	10
10	4	0
11	4	0

22.

Node		Weight	Pointer
1			5
2			7
3			8
4			9
5	2	3	6
6	3	1	0
7	4	2	0
8	4	2	0
9	1	1	0

23.

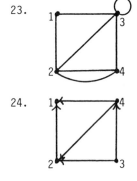

24.

*25.

	Left child	Right child
1	2	3
2	0	4
3	5	6
4	7	0
5	0	0
6	0	0
7	0	0

26.

	Left child	Right child
1	2	3
2	4	5
3	6	7
4	8	9
5	10	11
6	12	13
7	14	15
8	0	0
9	0	0
10	0	0
11	0	0
12	0	0
13	0	0
14	0	0
15	0	0

*27.

28.

29.

30.

*31. ρ = {(1,1), (1,2), (2,3), (3,4), (4,3), (4,2)}

32.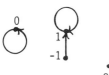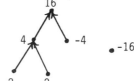

Section 4.3

*1. yes

2. yes

3. no

4. no

*5. $\begin{bmatrix} 0 & 1 & 1 & 1 & 1 & 0 \\ 1 & 0 & 0 & 0 & 0 & 1 \\ 1 & 0 & 0 & 0 & 0 & 1 \\ 1 & 0 & 0 & 0 & 0 & 1 \\ 1 & 0 & 0 & 0 & 0 & 1 \\ 0 & 1 & 1 & 1 & 1 & 0 \end{bmatrix}$

Odd after row 2 is 0.

6. $\begin{bmatrix} 0 & 0 & 0 & 1 & 1 & 0 \\ 0 & 0 & 0 & 1 & 1 & 1 \\ 0 & 0 & 0 & 0 & 1 & 1 \\ 1 & 1 & 0 & 0 & 1 & 0 \\ 1 & 1 & 1 & 1 & 0 & 1 \\ 0 & 1 & 1 & 0 & 1 & 0 \end{bmatrix}$

Odd after row 4 is 2.

7. $n - 7 \leq 140n + 25$ for all $n \geq 1$
    $140n + 25 \leq 200(n - 7)$ for all $n \geq 24$

8. If $n \leq c\sqrt{n}$ for $n \geq n_0 > 0$ then $n^2 \leq c^2 n$
    for $n \geq n_0$, or $n \leq c^2$ for $n \geq n_0$, but eventually
    $n > c^2$ because c^2 is constant.

10. Each odd vertex is the beginning or end of such a path, so there exist at least n such paths. If each pair of odd vertices is connected by a temporary arc, the resulting graph has no odd vertices and has an Euler cycle. The removal of the temporary arcs leaves $\leq$ n disjoint Euler paths which traverse the original graph. Therefore n paths are necessary and sufficient.

*11. no

12. yes

13. yes

14. no

*15. Begin at any node and take one of the arcs out from that node. Each time a new node is entered on an arc, there is exactly one unused arc on which to exit that node; because the arc is unused, it will lead to a new node or to the initial node. Upon return to the initial node, if all nodes have been used, we are done. If there is an unused node, because the graph is connected, there is an unused path from that node to a used node, which means the used node has degree $\geq$ 3, a contradiction.

16. No; using rooms as nodes (plus an outside node), and doorways as arcs, the resulting graph has 4 odd vertices, hence no Euler path.

*17.   IN = {2}

	1	3	4	5	6	7	8
d:	3	2	∞	∞	∞	1	∞
s:	2	2	2	2	2	2	2

p = 7
IN = {2,7}

	1	3	4	5	6	7	8
d:	3	2	∞	∞	6	1	2
s:	2	2	2	2	7	2	7

$$p = 3$$
$$IN = \{2,7,3\}$$

	1	3	4	5	6	7	8
d:	3	2	3	∞	6	1	2
s:	2	2	3	2	7	2	7

$$p = 8$$
$$IN = \{2,7,3,8\}$$

	1	3	4	5	6	7	8
d:	3	2	3	3	6	1	2
s:	2	2	3	8	7	2	7

$$p = 5$$
$$IN = \{2,7,3,8,5\}$$

	1	3	4	5	6	7	8
d:	3	2	3	3	6	1	2
s:	2	2	3	8	7	2	7

path: 2,7,8,5 length = 3

18. $$IN = \{3\}$$

	1	2	4	5	6	7	8
d:	5	2	1	∞	∞	∞	2
s:	3	3	3	3	3	3	3

$$p = 4$$
$$IN = \{3,4\}$$

	1	2	4	5	6	7	8
d:	5	2	1	5	∞	∞	2
s:	3	3	3	4	3	3	3

$$p = 2$$
$$IN = \{3,4,2\}$$

	1	2	4	5	6	7	8
d:	5	2	1	5	∞	4	2
s:	3	3	3	4	3	2	3

$$p = 8$$
$$IN = \{3,4,2,8\}$$

	1	2	4	5	6	7	8
d:	5	2	1	3	∞	3	2
s:	3	3	3	8	3	8	3

$$p = 5$$
$$IN = \{3,4,2,8,5\}$$

	1	2	4	5	6	7	8
d:	5	2	1	3	9	3	2
s:	3	3	3	8	5	8	3

$$p = 7$$
$$IN = \{3,4,2,8,5,7\}$$

	1	2	4	5	6	7	8
d:	5	2	1	3	8	3	2
s:	3	3	3	8	7	8	3

$$p = 1$$
$$IN = \{3,4,2,8,5,7,1\}$$

	2	4	5	6	7	8
d:	2	1	3	6	3	2
s:	3	3	8	1	8	3

$$p = 6$$
$$IN = \{3,4,2,8,5,7,1,6\}$$

	2	4	5	6	7	8
d:	2	1	3	6	3	2
s:	3	3	8	1	8	3

path: 3,1,6 length = 6

(alternate path: 3,2,1,6)

19.   IN = {1}

	2	3	4	5	6	7	8
d:	3	5	∞	8	1	∞	∞
s:	1	1	1	1	1	1	1

p = 6
IN = {1,6}

	2	3	4	5	6	7	8
d:	3	5	∞	7	1	6	∞
s:	1	1	1	6	1	6	1

p = 2
IN = {1,6,2}

	2	3	4	5	6	7	8
d:	3	5	∞	7	1	4	∞
s:	1	1	1	6	1	2	1

p = 7
IN = {1,6,2,7}

	2	3	4	5	6	7	8
d:	3	5	∞	7	1	4	5
s:	1	1	1	6	1	2	7

p = 3
IN = {1,6,2,7,3}

	2	3	4	5	6	7	8
d:	3	5	6	7	1	4	5
s:	1	1	3	6	1	2	7

p = 8
IN = {1,6,2,7,3,8}

	2	3	4	5	6	7	8
d:	3	5	6	6	1	4	5
s:	1	1	3	8	1	2	7

p = 5
IN = {1,6,2,7,3,8,5}

	2	3	4	5	6	7	8
d:	3	5	6	6	1	4	5
s:	1	1	3	8	1	2	7

path: 1,2,7,8,5   length = 6

20.   IN = {4}

	1	2	3	5	6	7	8
d:	∞	∞	1	4	∞	∞	∞
s:	4	4	4	4	4	4	4

p = 3
IN = {4,3}

	1	2	3	5	6	7	8
d:	6	3	1	4	∞	∞	3
s:	3	3	4	4	4	4	3

p = 2
IN = {4,3,2}

	1	2	3	5	6	7	8
d:	6	3	1	4	∞	4	3
s:	3	3	4	4	4	2	3

p = 8
IN = {4,3,2,8}

	1	2	3	5	6	7	8
d:	6	3	1	4	∞	4	3
s:	3	3	4	4	4	2	3

p = 7
IN = {4,3,2,8,7}

	1	2	3	5	6	7	8
d:	6	3	1	4	9	4	3
s:	3	3	4	4	7	2	3

path: 4,3,2,7   length = 4

*21.   IN = {1}

	2	3	4	5	6	7
d:	2	∞	∞	3	2	∞
s:	1	1	1	1	1	1

           p = 2
        IN = {1,2}

	2	3	4	5	6	7
d:	2	3	∞	3	2	∞
s:	1	2	1	1	1	1

           p = 6
       IN = {1,2,6}

	2	3	4	5	6	7
d:	2	3	∞	3	2	5
s:	1	2	1	1	1	6

           p = 3
      IN = {1,2,6,3}

	2	3	4	5	6	7
d:	2	3	4	3	2	5
s:	1	2	3	1	1	6

           p = 5
      IN = {1,2,6,3,5}

	2	3	4	5	6	7
d:	2	3	4	3	2	5
s:	1	2	3	1	1	6

           p = 4
      IN = {1,2,6,3,5,4}

	2	3	4	5	6	7
d:	2	3	4	3	2	5
s:	1	2	3	1	1	6

           p = 7
      IN = {1,2,6,3,5,4,7}

	2	3	4	5	6	7
d:	2	3	4	3	2	5
s:	1	2	3	1	1	6

path: 1,6,7   length = 5

22.   IN = {3}

	1	2	4	5	6	7
d:	∞	∞	1	∞	∞	∞
s:	3	3	3	3	3	3

           p = 4
      IN = {3,4}

	1	2	4	5	6	7
d:	∞	∞	1	2	∞	2
s:	3	3	3	4	3	4

           p = 5
      IN = {3,4,5}

	1	2	4	5	6	7
d:	∞	∞	1	2	3	2
s:	3	3	3	4	5	4

           p = 7
      IN = {3,4,5,7}

	1	2	4	5	6	7
d:	∞	∞	1	2	3	2
s:	3	3	3	4	5	4

           p = 6
      IN = {3,4,5,7,6}

	1	2	4	5	6	7
d:	∞	∞	1	2	3	2
s:	3	3	3	4	5	4

           p = 1
      IN = {3,4,5,7,6,1}

	1	2	4	5	6	7
d:	∞	∞	1	2	3	2
s:	3	3	3	4	5	4

No path from 3 to 1

*23. 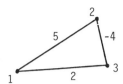 To find the shortest path from 1 to 3, the algorithm proceeds as follows:

$$I = \{1\}$$

d: $\begin{array}{cc} 2 & 3 \\ 5 & 2 \end{array}$

s: $\begin{array}{cc} 1 & 1 \end{array}$

$$p = 3$$
$$IN = \{1,3\}$$

d: $\begin{array}{cc} 2 & 3 \\ -2 & 2 \end{array}$

s: $\begin{array}{cc} 3 & 1 \end{array}$

Thus the algorithm will select the path 1-3 with length 2, although the shortest path is 1-2-3 with length 1. Allowing negative weights makes the greedy property insufficient for success, because a path with an initially high weight can later have its weight reduced by negative values, but this cannot be seen locally.

24. The shortest path from 1 to 5 is 1-5 with length 5. If the algorithm added the node closest to IN at each step, it would choose the path 1-2-3-4-5 with length 10.

26. Change line 3 to continue the loop until all nodes are in IN.

27. IN = {1,6,2,7,8,5,3,4}

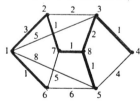

28. IN = {1,5,2,6,7,8,4,3}

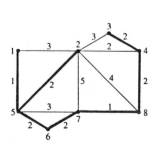

*29. IN = {1,8,5,6,2,7,4,3}

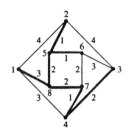

30. IN = {1,3,4,2,5,7,9,8,6}

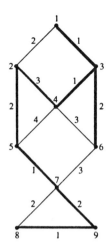

32. a) weight = 300   b) $d = 50\sqrt{2} = 70.7$
                          $4d = 282.8$

Section 4.4
*1. a b c e f d h g j i
 2. c a b e f d h g j i
 3. d a b c e f h g j i
 4. g b a c e f d h i j
*5. e b a c f d h g j i
 6. h f c a b e g j i d
*7. a b c f j g d e h k i
 8. e b a c f j g d h k i
*9. f c a b d e h k i g j
10. h e b a c f j g d i k
*11. a b c d e g f h j i
12. c a b e f d g h j i

84

13. d a f b c e h g i j
14. g b e h j a c f i d
15. e b c f g a d h j i
16. h f g i c d e b j a
*17. a b c d e f g h i j k
18. e b d h i a c k f g j
19. f c j a b g d e h i k
20. h e k b d i a c f g j
*21. preorder: a b d e h f c g
     inorder: d b h e f a g c
     postorder: d h e f b g c a
22. preorder: a b d g e c f h
     inorder: g d b e a h f c
     postorder: g d e b h f c a
23. preorder: a b e c f j g d h i
     inorder: e b a j f c g h d i
     postorder: e b j f g c h i d a
24. preorder: a b e f c g h d i
     inorder: e b f a g c h i d
     postorder: e f b g h c i d a
*25. preorder: a b c e f d g h
     inorder: e c f b g d h a
     postorder: e f c g h d b a
26. preorder: a b d h i c e f g
     inorder: h d i b a e c f g
     postorder: h i d b e f g c a
*27. prefix: + / 3 4 - 2 y
     postfix: 3 4 / 2 y - +

85

28. prefix: * + * x y / 3 z 4

    postfix: x y * 3 z / + 4 *

29. infix: ((2 + 3) * (6 * x)) - 7

    postfix: 2 3 + 6 x * * 7 -

30. infix: ((x - y) + z) - w

    postfix: x y - z + w -

*31. prefix: + * 4 - 7 x z

    infix: (4 * (7 - x)) + z

32. prefix: / x - + 2 w * y z

    infix: x / ((2 + w) - (y * z))

33.

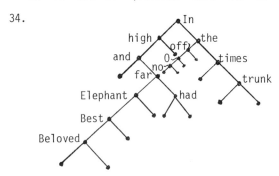

be is not or question that the To

34.

and Beloved Best Elephant far had high in no O off
the times trunk

*35.    both infix and postfix traversal give

d c b a

36.   both preorder traversals give a b c

37. Do a breadth-first search starting from the root and visiting adjacent nodes left to right.

CHAPTER 5

Section 5.1

*1.

+	0	1	a	a'
0	0	1	a	a'
1	1	1	1	1
a	a	1	a	1
a'	a'	1	1	a'

·	0	1	a	a'
0	0	0	0	0
1	0	1	a	a'
a	0	a	a	0
a'	0	a'	0	a'

2. a) $\max(x,y) = \max(y,x) \quad \min(x,y) = \min(y,x)$
    $\max(\max(x,y),z) = \max(x,\max(y,z)) = \max(x,y,z)$
    $\min(\min(x,y),z) = \min(x,\min(y,z)) = \min(x,y,z)$
    $\max(x,\min(y,z)) = \min(\max(x,y),\max(x,z))$
    $\min(x,\max(y,z)) = \max(\min(x,y),\min(x,z))$
    (The last two can be shown by taking cases:
    $x < y < z$, $y < x < z$, etc.)

    b) Let m be the 0 element. Then we must have $\max(x,m) = x$
    for all $x \in Z$. But for $x = m - 1$, $\max(m - 1, m) = m$.

3. a) $2^4 = 16$

    b) $(f_1 + f_2)(0,0) = 1 \quad (f_1 \cdot f_2)(0,0) = 1 \quad f_1'(0,0) = 0$
    $(f_1 + f_2)(0,1) = 1 \quad (f_1 \cdot f_2)(0,1) = 0 \quad f_1'(0,1) = 1$
    $(f_1 + f_2)(1,0) = 1 \quad (f_1 \cdot f_2)(1,0) = 0 \quad f_1'(1,0) = 0$
    $(f_1 + f_2)(1,1) = 0 \quad (f_1 \cdot f_2)(1,1) = 0 \quad f_1'(1,1) = 1$

    c) + and · are binary operations on B, ' is a unary operation
    on B. Max and min are cummutative and associative opera-
    tions; the distributive laws follow by considering different
    cases for the vaues of $f_1(x,y)$, $f_2(x,y)$, and $f_3(x,y)$ for
    a fixed $(x,y) \in S^2$. For any f in B and (x,y) in S^2,
    $(f + 0)(x,y) = \max(f(x,y), 0(x,y)) = \max(f(x,y), 0) = f(x,y)$,
    and $(f \cdot 1)(x,y) = \min(f(x,y), 1(x,y)) = \min(f(x,y), 1) = f(x,y)$.
    Also $(f + f')(x,y) = \max(f(x,y), f'(x,y)) = 1$ and $(f \cdot f')(x,y)$
    $= \min(f(x,y), f'(x,y)) = 0$ since one value of the pair
    $(f(x,y), f'(x,y))$ is 1 and the other 0.

4. *a) $x' + x = x + x'$ (1a)
    $\quad = 1$ (5a)
    and
    $x' \cdot x = x \cdot x'$ (1b)
    $\quad = 0$ (5b)
    Therefore $x = (x')'$ by Theorem 5.8.

*b) $x + (x \cdot y)$ = $x \cdot 1 + x \cdot y$ (4b)
                   = $x(1 + y)$ (3b)
                   = $x(y + 1)$ (1a)
                   = $x \cdot 1$ (Practice 5.7)
                   = x (4b)
        $x \cdot (x + y) = x$ follows by duality

c) $(x + y) + (x' \cdot y')$ = $((x + y) + x') \cdot ((x + y) + y')$ 3a
                         = $(x + (y + x')) \cdot (x + (y + y'))$ 2a
                         = $(x + (x' + y)) \cdot (x + 1)$ 1a,5a
                         = $((x + x') + y) \cdot 1$ 2a, Practice 5.7(a)
                         = $(1 + y) \cdot 1$ 5a
                         = $(y + 1) \cdot 1$ 1a
                         = $1 \cdot 1$ Practice 5.7(a)
and   = 1   4b

$(x + y) \cdot (x' \cdot y')$ = $(x' \cdot y') \cdot (x + y)$ 1b
                    = $(x' \cdot y') \cdot x + (x' \cdot y') \cdot y$ 3b
                    = $x \cdot (x' \cdot y') + (x' \cdot y') \cdot y$ 1b
                    = $(x \cdot x') \cdot y' + x' \cdot (y' \cdot y)$ 2b
                    = $0 \cdot y' + x' \cdot (y \cdot y')$ 5b,1b
                    = $y' \cdot 0 + x' \cdot 0$ 1b,5b
                    = $0 + 0$ Practice 5.7(b)
                    = 0   4a

Therefore $x' \cdot y' = (x + y)'$ by Theorem 5.8.
$(x \cdot y)' = x' + y'$ follows by duality.

d) $x \cdot [y + (x \cdot z)]$ = $x \cdot y + x \cdot (x \cdot z)$ 3b
                   = $x \cdot y + (x \cdot x) \cdot z$ 2b
                   = $x \cdot y + x \cdot z$ (dual of idempotent property)
   $x + [y \cdot (x + z)]$ = $(x + y) \cdot (x + z)$ follows by duality

e) $(x + y) \cdot (x' + y)$ = $(y + x) \cdot (y + x')$ 1a
                       = $y + (x \cdot x')$ 3a
                       = $y + 0$ 5b
                       = y 4a
   $(x \cdot y) + (x' \cdot y) = y$ by duality

f) $(x + y) + (y \cdot x')$ = $((x + y) + y) \cdot ((x + y) + x')$ 3a
                      = $(x + (y + y)) \cdot ((x + y) + x')$ 2a
                      = $(x + y) \cdot ((x + y) + x')$ idempotent
                      = $(x + y) \cdot (x' + (x + y))$ 1a

$$
\begin{aligned}
&= (x + y) \cdot ((x' + x) + y) &&\text{2a}\\
&= (x + y) \cdot ((x + x') + y) &&\text{1a}\\
&= (x + y) \cdot (1 + y) &&\text{5a}\\
&= (x + y) \cdot (y + 1) &&\text{1a}\\
&= (x + y) \cdot 1 &&\text{Practice 5.7(a)}\\
&= x + y &&\text{4b}
\end{aligned}
$$

$$(x \cdot y) \cdot (y + x') = x \cdot y \qquad \text{by duality}$$

g)
$$
\begin{aligned}
x + y' &= (x + 0) + y' &&\text{4a}\\
&= (x + x \cdot x') + y' &&\text{5b}\\
&= x + (x \cdot x' + y') &&\text{2a}\\
&= x + [((x \cdot x')')' + y'] &&\text{Exercise 4a}\\
&= x + [(x \cdot x')' \cdot y]' &&\text{De Morgan's Laws}\\
&= x + [(x' + (x')') \cdot y]' &&\text{De Morgan's Laws}\\
&= x + [(x' + x) \cdot y]' &&\text{Exercise 4(a)}\\
&= x + [y \cdot (x' + x)]' &&\text{1b}\\
&= x + [y \cdot x' + y \cdot x]' &&\text{3b}\\
&= x + (x' \cdot y + x \cdot y)' &&\text{1b}
\end{aligned}
$$

h)
$$
\begin{aligned}
((x \cdot y) \cdot z) + (y \cdot z) &= (x \cdot (y \cdot z)) + (y \cdot z) &&\text{2b}\\
&= ((y \cdot z) \cdot x) + y \cdot z &&\text{1b}\\
&= ((y \cdot z) \cdot x) + (y \cdot z) \cdot 1 &&\text{4b}\\
&= (y \cdot z) \cdot (x + 1) &&\text{3b}\\
&= (y \cdot z) \cdot 1 &&\text{Practice 5.7}\\
&= y \cdot z &&\text{4b}
\end{aligned}
$$

i)
$$
\begin{aligned}
(x + y) \cdot (y + 1) &= (x + y) \cdot y + (x + y) \cdot 1 &&\text{3b}\\
&= (x + y) \cdot y + (x + y) &&\text{4b}\\
&= y \cdot (x + y) + (x + y) &&\text{1b}\\
&= (y \cdot x + y \cdot y) + (x + y) &&\text{3b}\\
&= (x \cdot y + y \cdot y) + (x + y) &&\text{1b}\\
&= (x \cdot y + y) + (x + y) &&\text{dual of idempotent property}\\
&= (x \cdot y + y) + (y + x) &&\text{1a}\\
&= ((x \cdot y + y) + y) + x &&\text{2a}\\
&= (x \cdot y + (y + y)) + x &&\text{2a}\\
&= (x \cdot y + y) + x &&\text{idempotent property}\\
&= x + (x \cdot y + y) &&\text{1a}
\end{aligned}
$$

j) $((x' + z') \cdot (y + z'))' = (x' + z')' + (y + z')'$ De Morgan's Laws
$= ((x')' \cdot (z')') + y' \cdot (z')'$ De Morgan's Laws
$= x \cdot z + y' \cdot z$ Exercise 4a
$= z \cdot x + z \cdot y'$ 1b
$= z \cdot (x + y')$ 3b
$= (x + y') \cdot z$ 1b

5. *a) $x \oplus y = x \cdot y' + y \cdot x'$ definition of $\oplus$
$= y \cdot x' + x \cdot y'$ 1a
$= y \oplus x$ definition of $\oplus$

b) $x \oplus x = x \cdot x' + x \cdot x'$ definition of $\oplus$
$= 0 + 0$ 5b
$= 0$ 4a

c) $0 \oplus x = 0 \cdot x' + x \cdot 0'$ definition of $\oplus$
$= x' \cdot 0 + x \cdot 0'$ 1b
$= 0 + x \cdot 0'$ Practice 5.7(b)
$= 0 + x \cdot 1$ Practice 5.9
$= 0 + x$ 4b
$= x + 0$ 1a
$= x$ 4a

d) $1 \oplus x = 1 \cdot x' + x \cdot 1'$ definition of $\oplus$
$= x' \cdot 1 + x \cdot 1'$ 1b
$= x' + x \cdot 1'$ 4b
$= x' + x \cdot 0$ Practice 5.9
$= x' + 0$ Practice 5.7(b)
$= x'$ 4a

6. a)   Let $x + y = 0$. Then $x = x + 0$ 4a
$= x + (x + y)$ $x + y = 0$
$= (x + x) + y$ 2a
$= x + y$ idempotent property
$= 0$ $x + y = 0$

Similarly, if $x + y = 0$ then $y = 0$.

b)   Let $x = y$. Then
$x \cdot y' + y \cdot x' = x \cdot x' + x \cdot x'$ $x = y$
$= 0 + 0$ 5b
$= 0$ 4a

Let $x \cdot y' + y \cdot x' = 0$. Then $x \cdot y' = 0$ and $y \cdot x' = 0$ from part(a). Also,

$x = (y \cdot x) + (y' \cdot x)$	Exercise 4(e)
$= (y \cdot x) + (x \cdot y')$	1b
$= y \cdot x + 0$	$x \cdot y' = 0$
$= y \cdot x$	4a
$= x \cdot y$	1b
$= x \cdot y + 0$	4a
$= (x \cdot y) + (y \cdot x')$	$y \cdot x' = 0$
$= (x \cdot y) + (x' \cdot y)$	1b
$= y$	Exercise 4(e)

7. a) From Example 5.3, let $x = 1$, $y = 0$, $z = 1$.
Then $x + y = 1 + 0 = 1$
$x + z = 1 + 1 = 1$
but $y \neq z$

b) Let $x + y = x + z$ and $x' + y = x' + z$. Then

$y = y \cdot (y + x)$	Exercise 4(b)
$= y \cdot (x + y)$	1a
$= y \cdot (x + z)$	$x + y = x + z$
$= y \cdot x + y \cdot z$	3b
$= (y \cdot x + y \cdot z) + 0$	4a
$= (y \cdot x + y \cdot z) + x \cdot x'$	5b
$= (x \cdot y + y \cdot z) + x \cdot x'$	1b
$= x \cdot y + x \cdot x' + y \cdot z$	1a and 2a
$= x \cdot (y + x') + y \cdot z$	3b
$= x \cdot (x' + y) + y \cdot z$	1a
$= x \cdot (x' + z) + y \cdot z$	$x' + y = x' + z$
$= x \cdot x' + x \cdot z + y \cdot z$	3b
$= 0 + x \cdot z + y \cdot z$	5b
$= x \cdot z + y \cdot z$	1a and 4a
$= z \cdot x + z \cdot y$	1b
$= z \cdot (x + y)$	3b
$= z \cdot (x + z)$	$x + y = x + z$
$= z \cdot (z + x)$	1a
$= z$	Exercise 4(b)

8. Suppose $x + 0_1 = x$ for all $x \in B$. Then
$0 + 0_1 = 0$ and $0_1 + 0 = 0_1$ so
$0_1 = 0_1 + 0 = 0 + 0_1 = 0$ and $0_1 = 0$. Then $1 = 0'$, so
1 is unique by uniqueness of the complement, Theorem 5.8.

9. *a) i) If $x \leq y$ then $x \leq y$ and $x \leq x$ so x is a lower bound of x and y. If $w* \leq x$ and $w* \leq y$, then $w* \leq x$ so x is a greatest lower bound, and $x = x \cdot y$. If $x = x \cdot y$, then x is a greatest lower bound of x and y, so $x \leq y$.
   ii) Similar to (i).

b) i) Let $x + y = z$. Then z is a least upper bound of x and y, which is a least upper bound of y and x, so $z = y + x$.
   ii) Similar to (i).
   iii) Let $(x + y) + z = p$ and $x + (y + z) = q$. Then $y \leq x + y \leq p$ and $z \leq p$ so p is an upper bound for y and z; because $y + z$ is the least upper bound for y and z, $y + z \leq p$. Also $x \leq x + y \leq p$. Therefore p is an upper bound for x and $y + z$, and $q \leq p$ because q is the least upper bound for x and $y + z$. Similarly $p \leq q$, so $p = q$.
   iv) Similiar to (iii).

c) $x + 0 = x \longleftrightarrow 0 \leq x$, which is true because 0 is a least element.
   $x \cdot 1 = x \longleftrightarrow x \leq 1$, which is true because 1 is a greatest element.

d) (a) no - no least element
   (b) yes
   (c) yes
   (d) no - not distributive:
   $2 + (3 \cdot 4) = 2 + 1 = 2$
   $(2 + 3) \cdot (2 + 4) = 5 \cdot 5 = 5$
   Also, both 3 and 4 are complements of 2, so complements are not unique.

10. Suppose B has an odd number of elements. Arrange these distinct elements as follows: $0, 1, x_1, x_1', x_2, x_2', \ldots, x_{n-1}, x_{n-1}', x_n$. Then x_n' exists in B and must be an element of the list. If $x_n' = x_n$, then $x_n + x_n = x_n$ by idempotent property, but

$x_n + x_n = 1$ by (5a); thus $x_n = 1$, a contradiction. If $x_n' = x_i$ for $i < n$, then $(x_n')' = x_n = x_i'$, a previous element of the list; contradiction. If $x_n' = x_i'$ for $i < n$, then $(x_n')' = x_n = x_i$, a previous element of the list; again, a contradiction.

11. Compute
$$\left(\begin{bmatrix} a_{11} & a_{12} \\ a_{21} & a_{22} \end{bmatrix} \cdot \begin{bmatrix} b_{11} & b_{12} \\ b_{21} & b_{22} \end{bmatrix}\right) \cdot \begin{bmatrix} c_{11} & c_{12} \\ c_{21} & c_{22} \end{bmatrix} \text{ and}$$
$$\begin{bmatrix} a_{11} & a_{12} \\ a_{21} & a_{22} \end{bmatrix} \cdot \left(\begin{bmatrix} b_{11} & b_{12} \\ b_{21} & b_{22} \end{bmatrix} \cdot \begin{bmatrix} c_{11} & c_{12} \\ c_{21} & c_{22} \end{bmatrix}\right).$$

The results are equal by the distributive, commutative, and associative properties of Z under addition and multiplication.

12. Addition is a binary operation on $M_2(Z)$:
$$\begin{bmatrix} a_{11} & a_{12} \\ a_{21} & a_{22} \end{bmatrix} + \begin{bmatrix} b_{11} & b_{12} \\ b_{21} & b_{22} \end{bmatrix} = \begin{bmatrix} a_{11}+b_{11} & a_{12}+b_{12} \\ a_{21}+b_{21} & a_{22}+b_{22} \end{bmatrix} \in M_2(Z)$$

Addition is associative:
$$\left(\begin{bmatrix} a_{11} & a_{12} \\ a_{21} & a_{22} \end{bmatrix} + \begin{bmatrix} b_{11} & b_{12} \\ b_{21} & b_{22} \end{bmatrix}\right) + \begin{bmatrix} c_{11} & c_{12} \\ c_{21} & c_{22} \end{bmatrix}$$
$$= \begin{bmatrix} a_{11}+b_{11} & a_{12}+b_{12} \\ a_{21}+b_{21} & a_{22}+b_{22} \end{bmatrix} + \begin{bmatrix} c_{11} & c_{12} \\ c_{21} & c_{22} \end{bmatrix}$$
$$= \begin{bmatrix} (a_{11}+b_{11})+c_{11} & (a_{12}+b_{12})+c_{12} \\ (a_{21}+b_{21})+c_{21} & (a_{22}+b_{22})+c_{22} \end{bmatrix}$$
$$= \begin{bmatrix} a_{11}+(b_{11}+c_{11}) & a_{12}+(b_{12}+c_{12}) \\ a_{21}+(b_{21}+c_{21}) & a_{22}+(b_{22}+c_{22}) \end{bmatrix}$$
$$= \begin{bmatrix} a_{11} & a_{12} \\ a_{21} & a_{22} \end{bmatrix} + \begin{bmatrix} b_{11}+c_{11} & b_{12}+c_{12} \\ b_{21}+c_{21} & b_{22}+c_{22} \end{bmatrix} = \begin{bmatrix} a_{11} & a_{12} \\ a_{21} & a_{22} \end{bmatrix}$$
$$+ \left(\begin{bmatrix} b_{11} & b_{12} \\ b_{21} & b_{22} \end{bmatrix} + \begin{bmatrix} c_{11} & c_{12} \\ c_{21} & c_{22} \end{bmatrix}\right),$$

Addition is commutative:
$$\begin{bmatrix} a_{11} & a_{12} \\ a_{21} & a_{22} \end{bmatrix} + \begin{bmatrix} b_{11} & b_{12} \\ b_{21} & b_{22} \end{bmatrix} = \begin{bmatrix} a_{11}+b_{11} & a_{12}+b_{12} \\ a_{21}+b_{21} & a_{22}+b_{22} \end{bmatrix}$$
$$= \begin{bmatrix} b_{11}+a_{11} & b_{12}+a_{12} \\ b_{21}+a_{21} & b_{22}+a_{22} \end{bmatrix} = \begin{bmatrix} b_{11} & b_{12} \\ b_{21} & b_{22} \end{bmatrix} + \begin{bmatrix} a_{11} & a_{12} \\ a_{21} & a_{22} \end{bmatrix}$$

An identity exists:
$$\begin{bmatrix} 0 & 0 \\ 0 & 0 \end{bmatrix} + \begin{bmatrix} a_{11} & a_{12} \\ a_{21} & a_{22} \end{bmatrix} = \begin{bmatrix} a_{11} & a_{12} \\ a_{21} & a_{22} \end{bmatrix} = \begin{bmatrix} a_{11} & a_{12} \\ a_{21} & a_{22} \end{bmatrix} + \begin{bmatrix} 0 & 0 \\ 0 & 0 \end{bmatrix}$$

*13. a) $(x - y) - z = x - (y - z)$; no because, for example, $(2 - 3) - 4 \neq 2 - (3 - 4)$

b) $x - y = y - x$; no because, for example, $3 - 2 \neq 2 - 3$

14. *a) associative *b) commutative
    c) neither   d) commutative, associative
    e) commutative

Section 5.2

*1. a) $f(-2) = 2^{-2}$ and $f(12) = 2^{12}$. In $[R^+,\cdot]$,
    $2^{-2} \cdot 2^{12} = 2^{10}$. Then $f^{-1}(2^{10}) = 10$.
  b) $f^{-1}(4) = \log_2 4 = 2$ and $f^{-1}(8) = \log_2 8 = 3$. In
    $[R,+]$, $2 + 3 = 5$. Then $f(5) = 2^5 = 32$.

2. a) $f(1) = \{1, 2\}$ and $f(a') = \{2\}$. In the Boolean algebra
    on $\mathcal{P}(\{1, 2\})$, $\{1,2\} \cap \{2\} = \{2\}$. Then $f^{-1}(\{2\}) = a'$.
  b) $f(a) = \{1\}$ and in the Boolean algebra on $\mathcal{P}(\{1, 2\})$,
    $\{1\}' = \{2\}$. Then $f^{-1}(\{2\}) = a'$.
  c) $f^{-1}(\{1\}) = a$ and $f^{-1}(\{2\}) = a'$. In the Boolean algebra
    on B, $a + a' = 1$. Then $f(1) = \{1, 2\}$.
  d) $f^{-1}(\{1\}) = a$ and $f^{-1}(\{1, 2\}) = 1$. In the Boolean algebra
    on B, $a \cdot 1 = a$. Then $f(a) = \{1\}$.

*3. a)
    b)
    c)

4. For example,

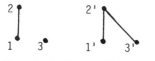

    $f(1) = 1'$
    $f(2) = 2'$
    $f(3) = 3'$

   Then $1 < 2$ and $f(1) <' f(2)$, but $f(3) <' f(2)$ and not $3 < 2$.

5. a) (i) bijection   (ii) for $x,y \in S$, $f(x \cdot y) = f(x) + f(y)$
  b) Let $f(0) = 5$, $f(1) = 7$. Then
    $f(0 \cdot 0) = f(1) = 7 = 5 + 5 = f(0) + f(0)$
    $f(0 \cdot 1) = f(0) = 5 = 5 + 7 = f(0) + f(1)$
    $f(1 \cdot 0) = f(0) = 5 = 7 + 5 = f(1) + f(0)$

$$f(1 \cdot 1) = f(1) = 7 = 7 + 7 = f(1) + f(1)$$

6. For example, $f_1 = f_4 + f_{13}$, therefore $f_1 \to \{1\} \cup \{3\} = \{1,3\}$. Therefore $f_1 \to \{1,3\}$. Similarly, $f_2 \to \{1,2\}$, $f_3 \to \{1,2,3\}$, $f_5 \to \{1,2,4\}$, $f_6 \to \{1,3,4\}$, $f_7 \to \{2,3,4\}$, $f_8 \to \{2,3\}$, $f_9 \to \{2,4\}$, $f_{10} \to \{1,4\}$, $f_{11} \to \{3,4\}$

*7. a) For any $y \in b$, $y = f(x)$ for some $x \in B$. Then $y \& f(0) = f(x) \& f(0) = f(x + 0) = f(x) = y$, and $f(0) = \emptyset$ because the zero element in any Boolean algebra is unique (see Exercise 8, Section 5.1).

   b) $f(1) = f(0') = [f(0)]'' = \emptyset'' = I$

8.
$$f\left(\begin{bmatrix} a_{11} & a_{12} \\ a_{21} & a_{22} \end{bmatrix} \cdot \begin{bmatrix} b_{11} & b_{12} \\ b_{21} & b_{22} \end{bmatrix}\right)$$
$$= f\left(\begin{bmatrix} a_{11}b_{11} + a_{12}b_{21} & a_{11}b_{12} + a_{12}b_{22} \\ a_{21}b_{11} + a_{22}b_{21} & a_{21}b_{12} + a_{22}b_{22} \end{bmatrix}\right)$$
$$= \begin{bmatrix} a_{11}b_{11} + a_{12}b_{21} & 0 \\ 0 & a_{21}b_{12} + a_{22}b_{22} \end{bmatrix}$$

but
$$f\left(\begin{bmatrix} a_{11} & a_{12} \\ a_{21} & a_{22} \end{bmatrix}\right) \cdot f\left(\begin{bmatrix} b_{11} & b_{12} \\ b_{21} & b_{22} \end{bmatrix}\right) = \begin{bmatrix} a_{11} & 0 \\ 0 & a_{22} \end{bmatrix} \cdot \begin{bmatrix} b_{11} & 0 \\ 0 & b_{22} \end{bmatrix}$$
$$= \begin{bmatrix} a_{11}b_{11} & 0 \\ 0 & a_{22}b_{22} \end{bmatrix}$$

9. a) R^* is closed under multiplication, multiplication is associative, and 1 is an identity.

   b) $f(a \cdot b) = |ab| = |a||b| = f(a) \cdot f(b)$

   c) For $a \in R^+$, $a = f(a)$, so f is onto. But $f(1) = f(-1)$, so f is not one-to-one.

   d) The class $[a] = \{a, -a\}$.

   e) Choose a member of the class associated with 3, say -3; choose a member of the class associated with 5, say 5. In $[R^*, \cdot], (-3)(5) = -15 \in [15]$. The corresponding member of $[R^+, \cdot]$ is 15.

   f) $f(-2) = |-2| = 2$, $f(6) = |6| = 6$. Then in $[R^+, \cdot]$, $2 \cdot 6 = 12$. The class associated with 12 is $[12] = \{12, -12\}$.

CHAPTER 6

Section 6.1

1. *a), b), c), d), 2.

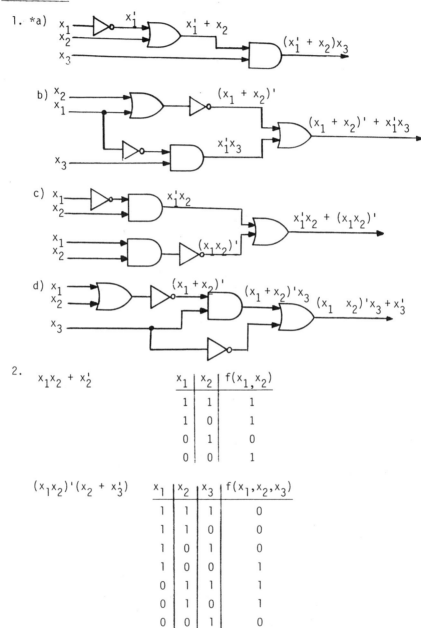

2. $x_1 x_2 + x_2'$

x_1	x_2	$f(x_1, x_2)$
1	1	1
1	0	1
0	1	0
0	0	1

$(x_1 x_2)'(x_2 + x_3')$

x_1	x_2	x_3	$f(x_1, x_2, x_3)$
1	1	1	0
1	1	0	0
1	0	1	0
1	0	0	1
0	1	1	1
0	1	0	1
0	0	1	0
0	0	0	1

97

$[(x_1' + x_2)x_3]'$

x_1	x_2	x_3	$f(x_1,x_2,x_3)$
1	1	1	0
1	1	0	1
1	0	1	1
1	0	0	1
0	1	1	0
0	1	0	1
0	0	1	0
0	0	0	1

$x_1(x_1 + x_2')(x_2x_3)'$

x_1	x_2	x_3	$f(x_1,x_2,x_3)$
1	1	1	0
1	1	0	1
1	0	1	1
1	0	0	1
0	1	1	0
0	1	0	0
0	0	1	0
0	0	0	0

3. a)

x	y	$f(x_1 y)$
1	1	0
1	0	1
0	1	1
0	0	0

b)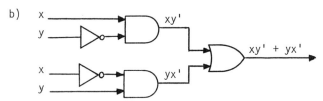

c) The truth function for the network of Figure 6.25 is also

x	y	$f(x_1 y)$
1	1	0
1	0	1
0	1	1
0	0	0

The network illustrates "x OR y" and "NOT both x AND y".

4. a) $x_1'x_2'$
   b) $x_1x_2 + x_1'x_2$
   c) $x_1x_2x_3' + x_1x_2'x_3 + x_1'x_2x_3 + x_1'x_2'x_3'$
   d) $x_1x_2'x_3 + x_1x_2'x_3' + x_1'x_2x_3'$

*5. a) $x_1x_2x_3' + x_1x_2'x_3'$

   b)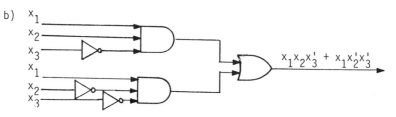

   c) $x_1x_2x_3' + x_1x_2'x_3' = x_1x_3'x_2 + x_1x_3'x_2' = x_1x_3'(x_2 + x_2')$
      $= x_1x_3' \cdot 1 = x_1x_3'$

   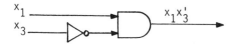

6. a) $x_1x_2x_3 + x_1'x_2x_3 + x_1'x_2x_3'$
   b)
   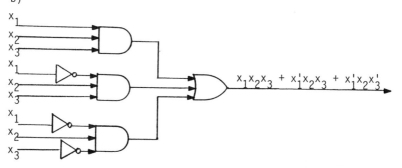

   c) $x_1x_2x_3 + x_1'x_2x_3 + x_1'x_2x_3' = x_2x_3x_1 + x_2x_3x_1' + x_1'x_2x_3' + x_1'x_2x_3$
      $= x_2x_3(x_1 + x_1') + x_1'x_2(x_3' + x_3)$
      $= x_2x_3 \cdot 1 + x_1'x_2 \cdot 1 = x_2x_3 + x_1'x_2$
      $= x_2x_3 + x_2x_1' + x_2(x_3 + x_1')$

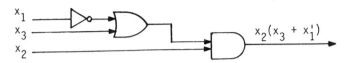

7. a)

x_1	x_2	x_3	$f(x_1,x_2,x_3)$
1	1	1	1
1	1	0	0
1	0	1	1
1	0	0	0
0	1	1	1
0	1	0	1
0	0	1	0
0	0	0	0

b) $x_1x_2x_3 + x_1x_2'x_3 + x_1'x_2x_3 + x_1'x_2x_3'$

c) $x_1x_3 + x_1'x_2 = (x_1x_3 + x_1')(x_1x_3 + x_2)$
$= (x_1' + x_1x_3)(x_2 + x_1x_3) = (x_1' + x_1)(x_1' + x_3)(x_2 + x_1)(x_2 + x_3)$
$= (x_1 + x_1')(x_1' + x_3)(x_1 + x_2)(x_2 + x_3)$
$= (x_1' + x_3)(x_1 + x_2)(x_2 + x_3) = (x_1 + x_2)(x_1' + x_3)(x_2 + x_3)$

*8. a) $(x_1' + x_2')(x_1' + x_2)(x_1 + x_2')$

b) $(x_1' + x_2)(x_1 + x_2)$

c) $(x_1' + x_2' + x_3')(x_1' + x_2 + x_3)(x_1 + x_2' + x_3)(x_1 + x_2 + x_3')$

d) $(x_1' + x_2' + x_3')(x_1' + x_2' + x_3)(x_1 + x_2' + x_3')(x_1 + x_2 + x_3')$
$(x_1 + x_2 + x_3)$

9.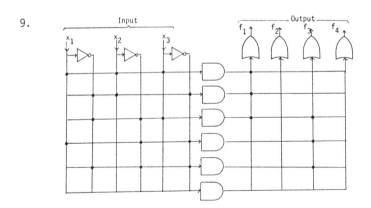

10. a) 1100   b) 1001   c) 011
       0100   0111   101
    (1)0000   (1)0000   (1)000

11. a)
x_i	c_{i-1}	r_i
0	0	0
0	1	1
1	0	1
1	1	0

If $c_{i-1} = 0$, no 1-digit has yet been seen, and $r_i = x_i$. If $c_{i-1} = 1$, a 1-digit has been seen and $r_i = x_i'$.

x_i	c_{i-1}	c_i
0	0	0
0	1	1
1	0	1
1	1	1

If $c_{i-1} = 1$, a 1-digit has already been seen, and $c_i = 1$ to convey this information to the next column. If $x_i = c_{i-1} = 0$, no 1 has been seen yet, and does not occur in column i, so $c_i = 0$. If $x_i = 1$ and $c_{i-1} = 0$, this is the first 1, so set $c_i = 1$.

b) $r_i = x_i' c_{i-1} + x_i c_{i-1}' = (x_i + c_{i-1})(x_i c_{i-1})'$

$c_i = x_i' c_{i-1} + x_i c_{i-1}' + x_i c_{i-1}$
$= x_i' c_{i-1} + x_i c_{i-1}' + x_i c_{i-1} + x_i c_{i-1}$
$= c_{i-1}(x_i' + x_i) + x_i(c_{i-1}' + c_{i-1})$
$= c_{i-1} + x_i$

c)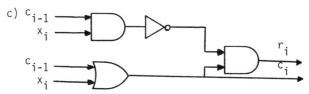

d)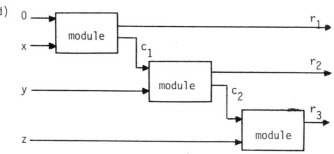

12. a)

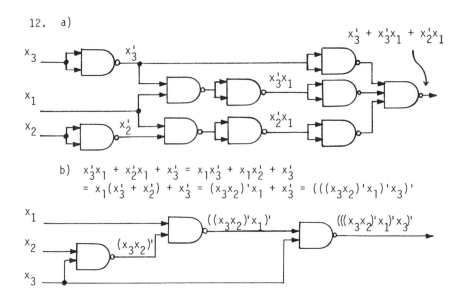

b) $x_3'x_1 + x_2'x_1 + x_3' = x_1x_3' + x_1x_2' + x_3'$
$= x_1(x_3' + x_2') + x_3' = (x_3x_2)'x_1 + x_3' = (((x_3x_2)'x_1)'x_3')'$

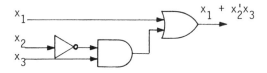

*13. Network is represented by $(x_1'(x_2'x_3)')'$ and
$(x_1'(x_2'x_3)')' = x_1 + x_2'x_3$

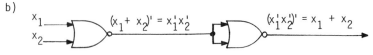

14. a)

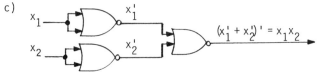

   $(x_1 + x_1)' = x_1'x_1' = x_1'$

   b) $(x_1 + x_2)' = x_1'x_2'$ $(x_1'x_2')' = x_1 + x_2$

   c) x_1' x_2' $(x_1' + x_2')' = x_1x_2$

*15. The truth function for | is that of the NAND gate, the truth function for ↓ is that of the NOR gate. In Section 1.1, we learned that every compound statement is equivalent to one using only |, or to one using

102

only ↓ , therefore any truth function can be realized by using only NAND gates or only NOR gates.

16. $\quad s = (x_1 x_2)(x_1 x_2)'$
$\quad\quad = x_1(x_1 x_2)' + x_2(x_1 x_2)'$
$\quad\quad = ((x_1(x_1 x_2)')' \cdot (x_2(x_1 x_2)')')'$
$\quad c = x_1 x_2 = ((x_1 x_2)')'$

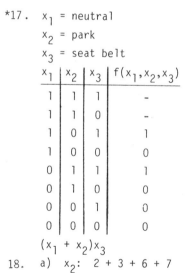

*17. x_1 = neutral
x_2 = park
x_3 = seat belt

x_1	x_2	x_3	$f(x_1,x_2,x_3)$
1	1	1	-
1	1	0	-
1	0	1	1
1	0	0	0
0	1	1	1
0	1	0	0
0	0	1	0
0	0	0	0

$(x_1 + x_2)x_3$

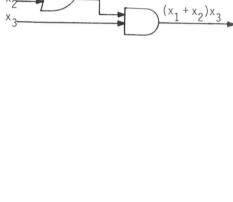

18. a) x_2: $2 + 3 + 6 + 7$

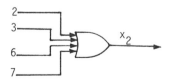

b)

	x_4	x_3	x_2	x_1	y_5	y_6
0	0	0	0	0	1	1
1	0	0	0	1	1	0
2	0	0	1	0	1	1
3	0	0	1	1	1	0
4	0	1	0	0	1	0
5	0	1	0	1	0	0
6	0	1	1	0	0	1
7	0	1	1	1	1	0
8	1	0	0	0	1	1
9	1	0	0	1	1	0

don't care

$$y_5 = (x_4 + x_3' + x_2 + x_1')(x_4 + x_3' + x_2' + x_1)$$
$$= (x_4 + x_3') + (x_2 + x_1')(x_2' + x_1)$$

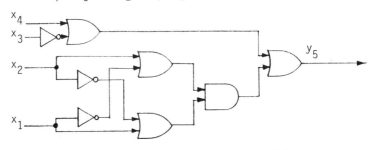

$$y_6 = x_4'x_3'x_2'x_1' + x_4'x_3'x_2x_1' + x_4'x_3x_2x_1' + x_4x_3'x_2'x_1'$$
$$= x_3'x_2'x_1' + x_4'x_2x_1'$$

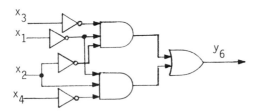

Section 6.2

1. *a)

	x_1x_2	x_1x_2'	$x_1'x_2'$	$x_1'x_2$
x_3			1	1
x_3'	1	1		1

$x_1'x_3 + x_1x_3' + x_1'x_2$

or

$x_1'x_3 + x_1x_3' + x_2x_3'$

104

b) 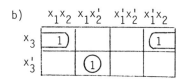 $x_2x_3 + x_1x_2'x_3'$

*c) $x_3 + x_2$

d) 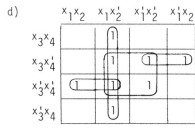 $x_1x_3'x_4' + x_1'x_3x_4' + x_2'x_4' + x_1x_2'$

e) 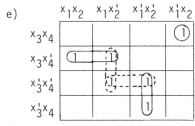 $x_1'x_2x_3x_4 + x_1x_3x_4' + x_1'x_2'x_3'$
$\qquad\qquad + x_1x_2'x_4'$

or

$x_1'x_2x_3x_4 + x_1x_3x_4' + x_1'x_2'x_3$
$\qquad\qquad + x_2'x_3'x_4'$

2. *a) 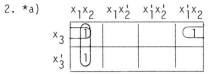 $x_1x_2 + x_2'x_3$

b) 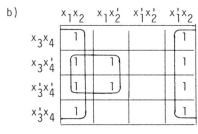 $x_1x_4' + x_2$

105

3. a)

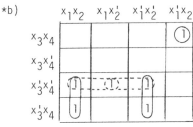

$x_2 x_3' x_4 + x_2' x_3 x_4' + x_1' x_4$

*b)

	$x_1 x_2$	$x_1 x_2'$	$x_1' x_2'$	$x_1' x_2$
$x_3 x_4$				1
$x_3 x_4'$				
$x_3' x_4'$	1		1	
$x_3' x_4$	1		1	

$x_1' x_2 x_3 x_4 + x_1' x_2 x_3' + x_1 x_2 x_3'$
$\quad + x_1 x_3' x_4'$

or

$x_1' x_2 x_3 x_4 + x_1' x_2 x_3' + x_1 x_2 x_3'$
$\quad + x_2 x_3' x_4'$

4. Original expression: $x_1 x_2 x_3 + x_1 x_2' x_3 + x_1 x_2' x_3' + x_1' x_2 x_3 + x_1' x_2 x_3'$

	$x_1 x_2$	$x_1 x_2'$	$x_1' x_2'$	$x_1' x_2$
x_3	1	1		1
x_3'		1		1

Reduced expression: $x_1 x_3 + x_1 x_2' + x_1' x_2$ or $x_2 x_3 + x_1 x_2' + x_1' x_2$

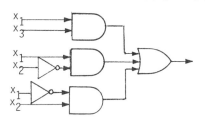

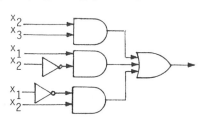

*5.

	$x_1 x_2$	$x_1 x_2'$	$x_1' x_2'$	$x_1' x_2$
$x_3 x_4$			1	
$x_3 x_4'$	1	–		1
$x_3' x_4'$	–			1
$x_3' x_4$			1	

$x_2 x_4' + x_1' x_2' x_4$

*6.

x_1	x_2	x_3	
1	1	1	1,2,3
1	0	1	1,4
0	1	1	2,5,6
1	1	0	3,7
0	0	1	4,5
0	1	0	6,7

x_1	x_2	x_3	
1	-	1	1
-	1	1	1,2
1	1	-	2
-	0	1	1
0	-	1	1
0	1	-	2
-	1	0	2

x_1	x_2	x_3	
-	-	1	
-	1	-	

	111	101	011	110	001	010
--1	✓	✓	✓		✓	
-1-	✓		✓	✓		✓

--1 and -1- are essential. The minimal form is $x_3 + x_2$.

7. Canonical sum-of-products form is

$x_1x_2x_3x_4 + x_1x_2'x_3x_4 + x_1'x_2x_3x_4' + x_1'x_2x_3x_4 + x_1x_2x_3'x_4 + x_1'x_2x_3'x_4$

x_1	x_2	x_3	x_4	
1	1	1	1	1,2,3
1	0	1	1	1,4
0	1	1	1	2,5
1	1	0	1	3,6
1	0	1	0	4
0	1	0	1	5,6

x_1	x_2	x_3	x_4	
1	-	1	1	
-	1	1	1	1
1	1	-	1	1
1	0	1	-	
0	1	-	1	1
-	1	0	1	1

x_1	x_2	x_3	x_4
-	1	-	1

	1111	1011	0111	1101	1010	0101
1-11	✓	✓				
101-		✓			✓	
-1-1	✓		✓	✓		✓

101- and -1-1 are essential; 1-11 is redundant. The minimal form is $x_1x_2'x_3 + x_2x_4$

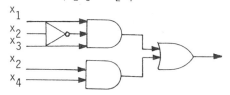

8. a)

x_1	x_2	x_3	x_4	
1	1	1	0	1
1	0	1	0	1,2,3
1	0	0	1	4
0	0	1	1	5
1	0	0	0	2,4,6
0	1	0	0	7
0	0	1	0	3,5,8
0	0	0	0	6,7,8

x_1	x_2	x_3	x_4	
1	-	1	0	
1	0	-	0	1
-	0	1	0	1
1	0	0	-	
0	0	1	-	
-	0	0	0	1
0	-	0	0	
0	0	-	0	1

x_1	x_2	x_3	x_4
-	0	-	0

	1110	1010	1001	0011	1000	0100	0010	0000
1-10	✓	✓						
100-			✓		✓			
001-				✓			✓	
0-00						✓		✓
-0-0		✓			✓		✓	✓

1-10, 100-, 001-, 0-00 are essential; -0-0 is redundant.

The minimal form is $x_1 x_3 x_4' + x_1 x_2' x_3' + x_1' x_2' x_3 + x_1' x_3' x_4'$

b)

x_1	x_2	x_3	x_4	
1	1	1	1	1,2,3
1	1	0	1	1,4,5
1	0	1	1	2,6,7
0	1	1	1	3,8,9
1	0	0	1	4,6,10,11
0	1	0	1	5,8,12,13
0	0	1	1	7,9,14
1	0	0	0	10,15
0	1	0	0	12,16
0	0	0	1	11,13,14,17
0	0	0	0	15,16,17

x_1	x_2	x_3	x_4	
1	1	-	1	1,2
1	-	1	1	1,3
-	1	1	1	2,3
1	-	0	1	1,4
-	1	0	1	2,4
1	0	-	1	1,5
-	0	1	1	3,5
0	1	-	1	2,6
0	-	1	1	3,6
1	0	0	-	7
-	0	0	1	4,5,7
0	1	0	-	8
0	-	0	1	4,6,8
0	0	-	1	5,6
-	0	0	0	7
0	-	0	0	8
0	0	0	-	7,8

$$\begin{array}{cccc} x_1 & x_2 & x_3 & x_4 \\ \hline 1 & - & - & 1 \\ - & 1 & - & 1 \\ - & - & 1 & 1 \\ - & - & 0 & 1 \\ - & 0 & - & 1 \\ 0 & - & - & 1 \\ - & 0 & 0 & - \\ 0 & - & 0 & - \end{array}^{\,1}_{\,1\,1\,1\,1\,1}$$

$$\begin{array}{cccc} x_1 & x_2 & x_3 & x_4 \\ \hline - & - & - & 1 \end{array}$$

	1111	1101	1011	0111	1001	0101	0011	1000	0100	0001	0000
-00-								✓		✓	
0-0-						✓			✓	✓	✓
---1	✓	✓	✓	✓	✓	✓	✓			✓	✓

-00-, 0-0-, and ---1 are essential. The minimal form is $x_2'x_3' + x_1'x_3' + x_4$

9. *a)
$$\begin{array}{cccc} x_1 & x_2 & x_3 & x_4 \\ \hline 0 & 1 & 1 & 1 \\ 1 & 0 & 1 & 0 \\ 0 & 0 & 1 & 1 \\ 0 & 1 & 0 & 0 \\ 0 & 0 & 0 & 1 \\ 0 & 0 & 0 & 0 \end{array}\begin{array}{l} 1 \\ \\ 1,2 \\ 3 \\ 2,4 \\ 3,4 \end{array}$$

$$\begin{array}{cccc} x_1 & x_2 & x_3 & x_4 \\ \hline 0 & - & 1 & 1 \\ 0 & 0 & - & 1 \\ 0 & - & 0 & 0 \\ 0 & 0 & 0 & - \end{array}$$

	0111	1010	0011	0100	0001	0000
1010		✓				
0-11	✓		✓			
00-1			✓		✓	
0-00				✓		✓
000-					✓	✓

1010, 0-11, 0-00 are essential. Either 00-1 or 000- can be used as the fourth term. The minimal sum-of-products form is $x_1x_2'x_3x_4' + x_1'x_3x_4 + x_1'x_3'x_4' + x_1'x_2'x_4$ or $x_1x_2'x_3x_4' + x_1'x_3x_4 + x_1'x_3'x_4' + x_1'x_2'x_3'$

109

b)

x_1 x_2 x_3 x_4	
1 1 1 1	1,2,3
1 0 1 1	1,4
1 1 1 0	2,5,6,7
1 1 0 1	3,8,9
1 0 1 0	4,5
0 1 1 0	6,10
1 1 0 0	7,8,11
0 1 0 1	9,12
0 1 0 0	10,11,12

x_1 x_2 x_3 x_4	
1 - 1 1	1
1 1 1 -	1,2
1 1 - 1	2
1 0 1 -	1
1 - 1 0	1
- 1 1 0	3
1 1 - 0	2,3
1 1 0 -	2,4
- 1 0 1	4
0 1 - 0	3
- 1 0 0	3,4
0 1 0 -	4

x_1 x_2 x_3 x_4
1 - 1 -
1 1 - -
- 1 - 0
- 1 0 -

	1111	1011	1110	1101	1010	0110	1100	0101	0100
1-1-	✓	✓	✓		✓				
11--	✓		✓	✓			✓		
-1-0			✓			✓	✓		✓
-10-						✓		✓	✓

1-1-, -1-0, -10- are essential; 11-- is redundant. The minimal form is $x_1 x_3 + x_2 x_4' + x_2 x_3'$

c)

x_1 x_2 x_3 x_4	
1 1 1 1	1,2
1 1 1 0	1,3,4
1 1 0 1	2,5,6
0 1 1 0	3,7
1 1 0 0	4,5,8
1 0 0 1	6,9
0 1 0 0	7,8,10
0 0 0 1	9,11
0 0 0 0	10,11

x_1 x_2 x_3 x_4	
1 1 1 -	1
1 1 - 1	1
- 1 1 0	2
1 1 - 0	1,2
1 1 0 -	1
1 - 0 1	
0 1 - 0	2
- 1 0 0	2
- 0 0 1	
0 - 0 0	
0 0 0 -	

x_1 x_2 x_3 x_4
1 1 - -
- 1 - 0

110

	1111	1110	1101	0110	1100	1001	0100	0001	0000
1-01			✓			✓			
-001						✓		✓	
0-00							✓		✓
000-								✓	✓
11--	✓	✓	✓		✓				
-1-0		✓		✓	✓		✓		

11--, -1-0 are essential. The additional terms can be 1-01 and 000-, or -001 and 0-00, or -001 and 000-. The minimal form is

$$x_1 x_2 + x_2 x_4' + x_1 x_3' x_4 + x_1' x_2' x_3'$$

or

$$x_1 x_2 + x_2 x_4' + x_2' x_3' x_4 + x_1' x_3' x_4'$$

or

$$x_1 x_2 + x_2 x_4' + x_2' x_3' x_4 + x_1' x_2' x_3'$$

d)

x_1	x_2	x_3	x_4	x_5	
1	1	1	1	1	1,2
1	0	1	1	1	1,3
1	1	0	1	1	2
0	1	1	0	1	4
1	0	1	0	1	3,5
1	1	1	0	0	
0	1	0	1	0	6
0	0	1	0	1	4,5,7
0	0	0	0	1	7

x_1	x_2	x_3	x_4	x_5
1	-	1	1	1
1	1	-	1	1
1	0	1	-	1
0	-	1	0	1
-	0	1	0	1
0	-	0	1	0
0	0	-	0	1

	11111	10111	11011	01101	10101	11100	01010	00101	00001	00010
11100						✓				
1-111	✓	✓								
11-11	✓		✓							
101-1		✓			✓					
0-101				✓				✓		
-0101				✓				✓		
0-010							✓			✓
00-01								✓	✓	

11100, 11-11, 0-101, 0-010, 00-01 are essential. The only single additional term that works is 101-1. The minimal form is

$$x_1x_2x_3x_4'x_5' + x_1x_2x_4x_5 + x_1'x_3x_4'x_5 + x_1'x_3x_4x_5' + x_1'x_2x_4'x_5$$
$$+ x_1x_2'x_3x_5$$

10.

x_1	x_2	x_3	x_4	
1	1	1	1	1,2
1	0	1	1	1,3
1	1	1	0	2,4
0	1	0	1	5
1	0	1	0	3,4,6
1	0	0	0	6,7
0	1	0	0	5,8
0	0	0	0	7,8

x_1	x_2	x_3	x_4	
1	-	1	1	1
1	1	1	-	1
1	0	1	-	1
1	-	1	0	1
0	1	0	-	
1	0	-	0	
-	0	0	0	
0	-	0	0	

x_1	x_2	x_3	x_4
1	-	1	-

	1111	1011	1110	0101	1010	1000	0100	0000
010-				✓			✓	
10-0					✓	✓		
-000						✓		✓
0-00							✓	✓
1-1-	✓	✓	✓		✓			

1-1- and 010- are essential. Either 10-0 and 0-00, or -000 will cover the remaining columns. Choosing -000, the minimal form is $x_1x_3 + x_1'x_2x_3' + x_2'x_3'x_4'$

CHAPTER 7

Section 7.1

*1. a) Not commutative: $a \cdot b \neq b \cdot a$
Not associative; $a \cdot (b \cdot d) \neq (a \cdot b) \cdot d$

b)

·	p	q	r	s
p	p	q	r	s
q	q	r	s	p
r	r	s	p	q
s	s	p	q	r

Commutative

2. For example:
   a) $a \cdot b = (a + b)^2$ b) $a \cdot b = a$
   c) $a \cdot b = a + 1$ d) $a \cdot b = a + b$

*3. a) f_0 = identity function
$f_1(1) = 2 \quad f_1(2) = 1$
$f_2(1) = 1 \quad f_2(2) = 1$
$f_3(1) = 2 \quad f_3(2) = 2$

∘	f_0	f_1	f_2	f_3
f_0	f_0	f_1	f_2	f_3
f_1	f_1	f_0	f_3	f_2
f_2	f_2	f_2	f_2	f_2
f_3	f_3	f_3	f_3	f_3

b) f_0 and f_1 are the elements

∘	f_0	f_1
f_0	f_0	f_1
f_1	f_1	f_0

4. *a) Semigroup
   *b) Not a semigroup - not associative
   *c) Not a semigroup - S not closed under·
   *d) Monoid; $i = 1 + 0\sqrt{2}$
   *e) Group; $i = 1 + 0\sqrt{2}$
   *f) Group; $i = 1$
   *g) Monoid; $i = 1$
   h) Monoid; $i = 1$

i) Semigroup
j) Monoid; i = (0,1)
k) Group; i = 0
l) Not a semigroup - S not closed under·
m) Group; i = zero polynomial
n) Not a semigroup = S not closed under·
o) Group; i = $\begin{bmatrix} 1 & 0 \\ 0 & 1 \end{bmatrix}$
p) Group; i = 1
q) Group; i = 0
r) Monoid; i = function mapping every x to 0

5. Points to include:
   i) all table entries must come from {1,2,...,10}
   ii) for all x,y,z ∈ {1,2,...,10}, (xy)z = x(yz)
   iii) must have a row which reads 1 2 ··· 10, if this is row i, then column i must read 1 2 ··· 10
   iv) each row must contain the $i^{\underline{th}}$ element (from step 3 above) and if i appears in location (m,n), then i must also appear in location (n,m).

6. a) i·i = i so i = i^{-1}
   b) $x^{-1} \cdot x = x \cdot x^{-1}$ = i so x = $(x^{-1})^{-1}$

7. a)

·	1	a
1	1	a
a	a	1

b)

·	1	a	b
1	1	a	b
a	a	b	1
b	b	1	a

c)

·	1	a	b	c
1	1	a	b	c
a	a	1	c	b
b	b	c	1	a
c	c	b	a	1

·	1	a	b	c
1	1	a	b	c
a	a	b	c	1
b	b	c	1	a
c	c	1	a	b

1↔1

·	1	a	b	c
1	1	a	b	c
a	a	c	1	b
b	b	1	c	a
c	c	b	a	1

a↔b
b↔c
c↔a

1↔1

·	1	a	b	c
1	1	a	b	c
a	a	1	c	b
b	b	c	a	1
c	c	b	1	a

a↔c
b↔b
c↔a

*8.

o	R_1	R_2	R_3	F_1	F_2	F_3
R_1	R_2	R_3	R_1	F_3	F_1	F_2
R_2	R_3	R_1	R_2	F_2	F_3	F_1
R_3	R_1	R_2	R_3	F_1	F_2	F_3
F_1	F_2	F_3	F_1	R_3	R_1	R_2
F_2	F_3	F_1	F_2	R_2	R_3	R_1
F_3	F_1	F_2	F_3	R_1	R_2	R_3

Identity element is R_3; inverse for F_1 is F_1; inverse for R_2 is R_1.

9. $\alpha_1 \rightarrow R_3$, $\alpha_2 \rightarrow F_3$, $\alpha_3 \rightarrow F_2$, $\alpha_4 \rightarrow F_1$, $\alpha_5 \rightarrow R_1$, $\alpha_6 \rightarrow R_2$.

10.

o	R_1	R_2	R_3	R_4	F_1	F_2	F_3	F_4
R_1	R_2	R_3	R_4	R_1	F_2	F_3	F_4	F_1
R_2	R_3	R_4	R_1	R_2	F_3	F_4	F_1	F_2
R_3	R_4	R_1	R_2	R_3	F_4	F_1	F_2	F_3
R_4	R_1	R_2	R_3	R_4	F_1	F_2	F_3	F_4
F_1	F_4	F_3	F_2	F_1	R_4	R_2	R_3	R_1
F_2	F_1	F_4	F_3	F_2	R_1	R_4	R_2	R_3
F_3	F_2	F_1	F_4	F_3	R_3	R_1	R_4	R_2
F_4	F_3	F_2	F_1	F_4	R_2	R_3	R_1	R_4

*11. a) $i_L = i_L \cdot i_R = i_R$ so $i_L = i_R$ and this element is an identity in $[S, \cdot]$.

b) For example,

·	a	b
a	a	b
b	a	b

c) For example,

·	a	b
a	a	a
b	b	b

d) For example, $[R^+, +]$.

12. a) $x_L^{-1} = x_L^{-1} \cdot i = x_L^{-1} \cdot (x \cdot x_R^{-1}) = (x_L^{-1} \cdot x) \cdot x_R^{-1} = i \cdot x_R^{-1} = x_R^{-1}$ so $x_L^{-1} = x_R^{-1}$ and this element is x^{-1}. Therefore every element has an inverse, and $[S, \cdot]$ is a group.

b) For $x \in N$, $(g \circ f)(x) = g(f(x)) = g(2x) = 2x/2 = x$. Therefore $g \circ f = i$, the identity function on N. If f had a right inverse, say h, then $(f \circ h)(x) = x$. But $(f \circ h)(x) = f(h(x)) = 2(h(x))$. For $2(h(x)) = x$ we must have $h(x) = x/2$, but this function is not a member of S.

c) For example, $[Z,\cdot]$

*13. a) $0_L = 0_L 0_R = 0_R$ b) For (a), zero is 0
For (d), zero is $0 + 0\sqrt{2}$
For (h), zero is 30

14. Supose x has a right inverse x_R^{-1}. Then $x \cdot x_R^{-1} = i$. If $|S| > 1$, then there is a $y \in S$, $y \neq i$, and $y = y \cdot i = y \cdot (x \cdot x_R^{-1}) = (y \cdot x) \cdot x_R^{-1} = x \cdot x_R^{-1}$ (since x is a right zero) $= i$. Contradiction.

15. $i \cdot i = i$, so there is one idempotent element. Let $y \in S$, and assume y is idempotent. Then $y \cdot y = y = y \cdot i$ and by cancellation, $y = i$.

16. Let $x \in G$. Then $x, x^2, x^3, \ldots$ are all members of G because G is closed under the operation. Because G is finite, these are not all distinct elements of G, and $x^n = x^m$ for some n and m, $m > n$. Then $\underbrace{x \cdot x \cdots x}_{n} \cdot i = \underbrace{x \cdot x \cdots x}_{m}$ and by left cancellation, $i = x^{m-n}$, $m-n > 0$.

*17. a) $x \rho x$ because $i \cdot x \cdot i^{-1} = x \cdot i^{-1} = x \cdot i = x$
If $x \rho y$ then for some $g \in G$, $g \cdot x \cdot g^{-1} = y$ or $g \cdot x = y \cdot g$ or $x = g^{-1} \cdot y \cdot g = (g^{-1}) \cdot y \cdot (g^{-1})^{-1}$ so $y \rho x$.
If $x \rho y$ and $y \rho z$ then for some $g_1, g_2 \in G$, $g_1 \cdot x \cdot g_1^{-1} = y$ and $g_2 \cdot y \cdot g_2^{-1} = z$ so
$g_2 \cdot g_1 \cdot x \cdot g_1^{-1} \cdot g_2^{-1} = z$ or $(g_2 \cdot g_1) \cdot x \cdot (g_2 \cdot g_1)^{-1} = z$ and $x \rho z$.

b) Suppose G is commutative and $y \in [x]$. Then for some $g \in G$, $y = g \cdot x \cdot g^{-1} = x \cdot g \cdot g^{-1} = x \cdot i = x$. Thus $[x] = \{x\}$. Conversely suppose $[x] = \{x\}$ for each $x \in G$, let $x, y \in G$, and denote the element $y \cdot x \cdot y^{-1}$ by z. Then $x \rho z$, so $z = x$ and $y \cdot x \cdot y^{-1} = x$ or $y \cdot x = x \cdot y$.

18. Let $x \in S$ with left inverse y. Then $y \in S$, so let z be the left inverse of y. Then $x \cdot y = i_L \cdot (x \cdot y) = (z \cdot y) \cdot (x \cdot y) = z \cdot (y \cdot x) \cdot y = z \cdot i_L \cdot y = z \cdot y = i_L$ so y is also a right inverse of x. Also, $x \cdot i_L = x \cdot (y \cdot x) = (x \cdot y) \cdot x = i_L \cdot x = x$, so i_L is also a right identity in S, therefore an identity.

19. For some fixed $a \in S$, let x_1 be the solution to $x \cdot a = a$. Let b be any element of S. Then $a \cdot x = b$ for some $x \in S$ and $x_1 \cdot b = x_1 \cdot (a \cdot x) = (x_1 \cdot a) \cdot x = a \cdot x = b$. Therefore x_1 is a left

identity in S. Also for any $b \in S$, there is an x such that $x \cdot b = x_1$, hence every element of S has a left inverse. Result follows from Exercise 18.

20. Let $S = \{x_1,\ldots,x_n\}$. The products $x_1 \cdot x_1, x_1 \cdot x_2, \ldots, x_1 \cdot x_n$ are distinct, for if $x_1 \cdot x_i = x_1 \cdot x_j$ then $x_i = x_j$ by left cancellation. These products are the n elements of S, so $x_1 \cdot x_i = x_1$ for some i. Then for any $x_j \in S$, $x_1 \cdot x_j = x_1 \cdot x_j$ $\rightarrow (x_1 \cdot x_i) \cdot x_j = x_1 \cdot x_j \rightarrow x_1 \cdot (x_i \cdot x_j) = x_1 \cdot x_j \rightarrow x_i \cdot x_j = x_j$ by left cancellation, and x_i is a left identity. Also, for any $x_j \in S$, form the n products $x_1 \cdot x_j, x_2 \cdot x_j, \ldots, x_n \cdot x_j$. These are distinct, by right cancellation, so $x_k \cdot x_j = x_i$ for some k, and x_k is thus a left inverse of x_j. Result follows from Exercise 18.

21. If G is commutative, then $(x \cdot y)^2 = (x \cdot y) \cdot (x \cdot y) = x \cdot (y \cdot x) \cdot y = x \cdot (x \cdot y) \cdot y = (x \cdot x) \cdot (y \cdot y) = x^2 \cdot y^2$. For the converse, let $x,y \in G$; then $x \cdot y \cdot x \cdot y = x \cdot x \cdot y \cdot y$, and by left and right cancellation, $y \cdot x = x \cdot y$, so G is commutative.

22. Let $x,y \in G$. Then $x \cdot y \cdot x \cdot y = i = x \cdot x \cdot y \cdot y$ and by left and right cancellation, $y \cdot x = x \cdot y$.

23. Suppose that for $x \in Z_n$, $x \neq 0$, there exists an inverse for x. If n is not prime, then $n = r \cdot s$ where $1 < r < n$ and $1 < s < n$. Then $r \cdot_n s = 0$. Let y be the inverse of r. Then $r \cdot_n y \cdot_n s = 1 \cdot_n s = s$. But also $r \cdot_n y \cdot_n s = r \cdot_n s \cdot_n y = 0 \cdot_n y = 0$. Hence $s = 0$, contradiction, and n is prime. Now suppose that n is prime and $x \in Z_n$, $x \neq 0$. Because n is prime, x and n are relatively prime and there exist integers a and b such that $ax + bn = 1$. If $0 < a < n$, then $a \in Z_n$ and $ax = (-b)n + 1$, so $a \cdot_n x = 1$ and $a = x^{-1}$. If $a \notin Z_n$, then a is not a multiple of n, otherwise $ax + bn$ would be a multiple of n that equals 1, which is not possible. Dividing a by n, $a = qn + r$, $0 < r < n$, so $r \in Z_n$. Then $rx = (a - qn)x = ax - qnx = 1 - bn - qnx = (-b - qx)n + 1$ so $r \cdot_n x = 1$, and $r = x^{-1}$.

Section 7.2

1. $\lambda \varepsilon$ set, and set is closed under concatenation.
2. *a) No - not the same operation
   *b) No - zero polynomial (identity) does not belong to P
   *c) No - not every element of Z^* has an inverse in Z^*
   d) Yes
   e) No - Z is not a subset of $M_2(Z)$
   f) Yes
   g) No - $\{0,3,6\}$ not closed under $+_8$
*3. $[\{0\},+_{12}]$, $[Z_{12},+_{12}]$, $[\{0,2,4,6,8,10\},+_{12}]$, $[\{0,4,8\},+_{12}]$
    $[\{0,3,6,9\},+_{12}]$, $[\{0,6\},+_{12}]$
4. R_4 is the identity; each set is closed; R_1 and R_3 are inverses of each other, and R_2, F_1, F_2, F_3, F_4 are self-inverses.
5. In each set closure holds, i is a member, and each element has an inverse.
6. a) R is nonempty and closed under concatenation.
   b) The submonoid consists of λ and all strings of 0's with any 1's separated by at least two 0's and no 1's occurring in the first two positions.
*7. $4!/2 = 24/2 = 12$ elements
   $\alpha_1 = i$ $\alpha_2 = (1,2)\circ(3,4)$ $\alpha_3 = (1,3)\circ(2,4)$ $\alpha_4 = (1,4)\circ(2,3)$
   $\alpha_5 = (1,3)\circ(1,2)$ $\alpha_6 = (1,2)\circ(1,3)$ $\alpha_7 = (1,3)\circ(1,4)$ $\alpha_8 = (1,4)\circ(1,2)$
   $\alpha_9 = (1,4)\circ(1,3)$ $\alpha_{10} = (1,2)\circ(1,4)$ $\alpha_{11} = (2,4)\circ(2,3)$ $\alpha_{12} = (2,3)\circ(2,4)$
8. a) $f, g: A \to A$. If $f(x_1) = f(x_2)$ then $1 - x_1 = 1 - x_2$ and $x_1 = x_2$, so f is one-to-one. If $g(x_1) = g(x_2)$ then $1/x_1 = 1/x_2$ and $x_1 = x_2$, so g is one-to-one. Also, f is onto since for $r \varepsilon A$, $1 - r \varepsilon A$ and $f(1 - r) = r$; g is onto since for $r \varepsilon A$, $1/r \varepsilon A$ and $g(1/r) = r$.
   b) The group consists of the identity function i, f, g, $f \circ g = 1 - \frac{1}{x}$, $g \circ f = \frac{1}{1-x}$, and $f \circ g \circ f = \frac{-x}{1-x}$.

o	i	f	g	fog	gof	fogof
i	i	f	g	fog	gof	fogof
f	f	i	fog	g	fogof	gof
g	g	gof	i	fogof	f	fog
fog	fog	fogof	f	gof	i	g
gof	gof	g	fogof	i	fog	f
fogof	fogof	fog	gof	f	g	i

*9. a) $S \cap T \subseteq G$. Closure: for $x,y \in S \cap T$, $x \cdot y \in S$ because of closure in S, $x \cdot y \in T$ because of closure in T, so $x \cdot y \in S \cap T$. Identity: $i \in S$ and $i \in T$ so $i \in S \cap T$. Inverses: for $x \in S \cap T$, $x^{-1} \in S$ and $x^{-1} \in T$ so $x^{-1} \in S \cap T$.

b) No. For example, $[\{0,4,8\},+_{12}]$ and $[\{0,6\},+_{12}]$ are subgroup of $[Z_{12},+_{12}]$ but $[\{0,4,6,8\},+_{12}]$ is not a subgroup of $[Z_{12},+_{12}]$ (not closed).

*10. Let $[S,\cdot]$ be a commutative monoid, let $A = \{a \mid a \in S, a^2 = a\}$. Then $A \subseteq S$; $i \cdot i = i$, so $i \in A$.
For $x,y \in A$, $(x \cdot y)^2 = (x \cdot y)(x \cdot y) = x \cdot x \cdot y \cdot y = x \cdot y$, so $x \cdot y \in A$.

11. Closure: let $s_1 \cdot t_1$, $s_2 \cdot t_2 \in ST$. Then $(s_1 \cdot t_1) \cdot (s_2 \cdot t_2) = s_1 \cdot s_2 \cdot t_1 \cdot t_2$, and $s_1 \cdot s_2 \in S$, $t_1 \cdot t_2 \in T$, so $s_1 \cdot s_2 \cdot t_1 \cdot t_2 \in ST$. Identity: $i \in S$ and $i \in T$ so $i = i \cdot i \in ST$. Inverses: let $s \cdot t \in ST$. Then $s^{-1} \in S$, $t^{-1} \in T$ so $s^{-1} \cdot t^{-1} \in ST$ and $s \cdot t \cdot s^{-1} \cdot t^{-1} = s \cdot s^{-1} \cdot t \cdot t^{-1} = i$, also $s^{-1} \cdot t^{-1} \cdot s \cdot t = i$.

*12. Closure: let $x,y \in B_k$. Then $(x \cdot y)^k = x^k \cdot y^k = i \cdot i = i$, so $x \cdot y \in B_k$. Identity: $i^k = i$, so $i \in B_k$. Inverses: for $x \in B_k$, $(x^{-1})^k = (x^k)^{-1} = i^{-1} = i$, so $x^{-1} \in B_k$.

13. Closure: let $x,y \in A$ with $x^n = i$, $y^m = i$. Then $(x \cdot y)^{nm} = x^{nm} \cdot y^{nm} = (x^n)^m \cdot (y^m)^n = i^m \cdot i^n = i$, so $x \cdot y \in A$.
Identity: $i^1 = i$, so $i \in A$.
Inverses: for $x \in A$, $x^n = i$, $(x^{-1})^n = (x^n)^{-1} = i^{-1} = i$, so $x^{-1} \in A$.

14. a) Closure: let $x,y \in A$. Then for any $g \in G$, $(x \cdot y) \cdot g = x \cdot (y \cdot g) = x \cdot (g \cdot y) = (x \cdot g) \cdot y = (g \cdot x) \cdot y = g \cdot (x \cdot y)$.
Identity: $i \cdot g = g \cdot i = g$ for all $g \in G$. Inverses: let $x \in A$. Then for any $g \in G$, $x \cdot g = g \cdot x$. Multiplying twice by x^{-1}, $g = x^{-1} \cdot g \cdot x$ and $g \cdot x^{-1} = x^{-1} \cdot g$, so $x^{-1} \in A$.

b) $\{R_3\}$

c) Let $G = A$; clearly G is then commutative. Now let G be commutative. We have $A \subseteq G$. Let $x \in G$. Then for any $g \in G$, $x \cdot g = g \cdot x$ so $x \in A$ and $G \subseteq A$.

d) $y \in G$ so $x \cdot y^{-1} \cdot y = y \cdot x \cdot y^{-1}$ or $x = y \cdot x \cdot y^{-1}$ or $x \cdot y = y \cdot x$.

*15. Closure: let (x,x) and (y,y) belong to A. Then $(x,x) \cdot (y,y) = (x \cdot y, x \cdot y) \in A$. Identity: $(i,i) \in A$. Inverses: for $(x,x) \in A$, $(x,x)^{-1} = (x^{-1}, x^{-1}) \in A$.

16. a) $\{\begin{bmatrix} 0 & a \\ a & 0 \end{bmatrix} | a \in Z, a > 0\}$ is a nonempty subset of the group $[M_2(Z), +]$. It is closed under addition, so it is a sub-semigroup, but the identity $\begin{bmatrix} 0 & 0 \\ 0 & 0 \end{bmatrix}$ is not a member, so it is not a subgroup.

b) Let $s \in S$. By closure of S, $s, s^2, s^3, \ldots$ all belong to S. Because G is finite, these are not all distinct elements, and $s^i = s^j$ for some i,j, $j > i$. By cancellation in G, $i = s^{j-i}$ and $i \in S$. Also, s^{j-i-1} is a nonnegative power of s, so $s^{j-i-1} \in S$ and $s \cdot s^{j-i-1} = s^{j-i-1} \cdot s = s^{j-i} = i$ so $s^{j-i-1} = s^{-1}$.

17. a) Closure: let $f, g \in H_a$. Then $(f \circ g)(a) = f(g(a)) = f(a) = a$, so $f \circ g \in H_a$. Identity: the identity mapping on A leaves a fixed. Inverses: let $f \in H_a$. Then $f(a) = a$ so $f^{-1}(a) = a$, and $f^{-1} \in H_a$.

b) $(n-1)!$

18. a) Identity: Let $x \in A$ ($A \neq \phi$). Then $x \cdot x^{-1} = i \in A$. Inverses: let $x \in A$. Then $i \cdot x^{-1} = x^{-1} \in A$. Closure: let $x, y \in A$. Then $y^{-1} \in A$, and $x \cdot (y^{-1})^{-1} = x \cdot y \in A$.

b) $B_k \neq \phi$ since $i \in B_k$. Let $x, y \in B_k$. Then $(x \cdot y^{-1})^k = x^k \cdot (y^{-1})^k = i \cdot (y^k)^{-1} = i \cdot i^{-1} = i$, so $x \cdot y^{-1} \in B_k$.

19. a) Let $x = a^{z_1}$, $y = a^{z_2} \in A$. Then $x \cdot y^{-1} = a^{z_1} \cdot (a^{z_2})^{-1} = a^{z_1} \cdot (a^{-1})^{z_2} = a^{z_1 - z_2} \in A$. By Exercise 18, A is a subgroup.

b) $2^0 = 0$, $2^1 = 2$, $2^2 = 2 +_7 2 = 4$, $2^3 = 6$, $2^4 = 1$, $2^5 = 3$, $2^6 = 5$.

c) $5^0 = 0$, $5^1 = 5$, $5^2 = 5 +_7 5 = 3$, $5^3 = 1$, $5^4 = 6$, $5^5 = 4$, $5^6 = 2$.

d) $2^0 = 1$, $2^1 = 2$, $2^2 = 2 \cdot_{11} 2 = 4$, $2^3 = 8$, $2^4 = 5$, $2^5 = 10$, $2^6 = 9$, $2^7 = 7$, $2^8 = 3$, $2^9 = 6$.

e) $3^0 = 0$, $3^1 = 3$, $3^2 = 3 +_4 3 = 2$, $3^3 = 1$

20. Let $x = a^{z_1}$, $y = a^{z_2}$ ε G. Then $x \cdot y = a^{z_1} \cdot a^{z_2} = a^{z_1 + z_2} = a^{z_2 + z_1} = a^{z_2} \cdot a^{z_1} = y \cdot x$.

21. Let a be the generator of G, and let [S,·] be a subgroup of G. If S = {i}, then S is cyclic with generator i. Let $x \varepsilon S$, $x \neq i$. Then $x = a^k$, $k \varepsilon Z$, $k \neq 0$. Either $k > 0$ or $a^{-k} = (a^k)^{-1} \varepsilon S$ and $-k > 0$; thus a to a positive integral power belongs to S. Let n be the smallest positive integer such that $a^n \varepsilon S$. Then a^n generates S because $\{(a^n)^z | z \varepsilon Z\} \subseteq S$ and vice versa. $\{(a^n)^z | z \varepsilon Z\} \subseteq S$ because $a^n \varepsilon S$ and S is closed. For the other direction, let $s \varepsilon S$, say $s = a^j$. Divide j by n, and then $j = qn + r$, $0 \leq r < n$. Then $a^j = a^{qn+r} = a^{qn} \cdot a^r$ or $(a^{qn})^{-1} \cdot a^j = a^r$. Then $a^{qn} = (a^n)^q \varepsilon S$ because of the closure of S, so $(a^{qn})^{-1} \varepsilon S$ and $(a^{qn})^{-1} \cdot a^j = a^r \varepsilon S$. Because n is the smallest positive integer such that $a^n \varepsilon S$, $r = 0$ and $j = qn$. Therefore $s = a^j = a^{qn} = (a^n)^q$ and $S \subseteq \{(a^n)^z | z \varepsilon Z\}$.

Section 7.3

1. *a) No - $f(x + y) = 2$, $f(x) + f(y) = 2 + 2 = 4$

   *b) No - $f(x + y) = x + y + 1$, $f(x) + f(y) = x + 1 + y + 1$

   *c) Yes - $f((x,y) + (p,q)) = f((x+p, y+q)) = x + p + 2(y + q)$,
   $f((x,y)) + f((p,q)) = x + 2y + p + 2q$

   *d) No - $f(x + y) = |x + y|$, $f(x) + f(y) = |x| + |y|$

   *e) Yes - $f(x \cdot y) = |x \cdot y|$, $f(x) \cdot f(y) = |x| \cdot |y|$

   f) Yes - $f(a_n x^n + \cdots + a_1 x + a_0 + b_k x^k + \cdots + b_1 x + b_0)$
   $= f(a_n x^n + a_{n-1} x^{n-1} + \cdots + (a_k + b_k) x^k + \cdots$
   $+ (a_1 + b_1) x + (a_0 + b_0))$
   $= a_n + a_{n-1} + \cdots + a_k + b_k + \cdots + a_1 + b_1 + a_0 + b_0$
   $= f(a_n x^n + \cdots + a_0) + f(b_k x^k + \cdots + b_0)$

   g) No - $f\left(\begin{bmatrix} a & 0 \\ 0 & b \end{bmatrix} \begin{bmatrix} c & 0 \\ 0 & d \end{bmatrix}\right) = f\left(\begin{bmatrix} ac & 0 \\ 0 & bd \end{bmatrix}\right) = (ac, bd)$

   $f\left(\begin{bmatrix} a & 0 \\ 0 & b \end{bmatrix}\right) + f\left(\begin{bmatrix} c & 0 \\ 0 & d \end{bmatrix}\right) = (a,b) + (c,d) = (a+c, b+d)$

h) No - take α and β both odd. Then $f(\alpha \circ \beta) = 1$ but $f(\alpha) +_2 f(\beta) = 0 +_2 0 = 0$.

None are isomorphisms.

2. a) Yes; $f: Z \rightarrow 12Z$, $f(x) = 12x$
   b) No; Z_5 is finite, $5Z$ is infinite
   c) Yes; $f: 5Z \rightarrow 12Z$, $f(5x) = 12x$
   d) No; $[S_3, \circ]$ is noncommutative, $[Z_6, +_6]$ is commutative
   e) Yes; $f: \{a_1 x + a_0\} \rightarrow C$, $f(a_1 x + a_0) = a_0 + a_1 i$
   f) No; the order of $[Z_6, +_6]$ is 6 but the order of $[S_6, \circ]$ is 6!
   g) Yes; $f: Z_2 \rightarrow S_2$, $f(x) = \begin{cases} i & \text{if } x = 0 \\ (1,2) & \text{if } x = 1 \end{cases}$

*3. a) $f: R^+ \rightarrow R$
   f is onto: for $r \in R$, $b^r \in R^+$ and $f(b^r) = \log_b b^r = r$
   f is one-to-one: if $f(x_1) = f(x_2)$, then $\log_b x_1 = \log_b x_2$. Let $p = \log_b x_1 = \log_b x_2$. Then $b^p = x_1$ and $b^p = x_2$ so $x_1 = x_2$.
   f is a homomorphism: for $x_1, x_2 \in R^+$, $f(x_1 \cdot x_2) = \log_b(x_1 \cdot x_2) = \log_b x_1 + \log_b x_2 = f(x_1) + f(x_2)$.
   b) $f(64) = \log_2 64 = 6$ and $f(512) = \log_2 512 = 9$. In $[R, +]$, $6 + 9 = 15$, and $f^{-1}(15) = 2^{15} = 32768$.

4. a) $\{0, 4, 8\}$
   b) $[0] = \{0, 3, 6, 9, 12, 15\}$
      $[1] = \{1, 4, 7, 10, 13, 16\}$
      $[2] = \{2, 5, 8, 11, 14, 17\}$
   c) The class associated with 4 is [1] and the class associated with 8 is [2]. To add $4 +_{12} 8$, choose an element of each class, say 1 and 2, add them in Z_{18} ($1 +_{18} 2 = 3$), then $3 \in [0]$ which corresponds to 0 in $f(Z_{18})$. Thus $4 +_{12} 8 = 0$.
   d) $f(14) = 8$ and $f(8) = 8$, so we add $8 +_{12} 8 = 4$. The class associated with 4 is [1], which contains the correct answer of 4.

5. a) $f(\alpha \cdot \beta) =$ the number of 1's in $\alpha\beta =$ the number of 1's in α plus the number of 1's in $\beta = f(\alpha) + f(\beta)$.
   b) Given $n \in N$, the string α of n 1's belongs to A^* and $f(\alpha) = n$.
   c) $f(\lambda) =$ the number of 1's in $\lambda = 0$
   d) $[\alpha] = \{$all strings of 0's and 1's with the same number of 1's as $\alpha\}$

e) The class associated with 3 is [111] and the class associated with 4 is [1111]. Then $111 \cdot 1111 = 1111111 \in$ [1111111], which corresponds to 7 in N.

f) $f(101) = 2$ and $f(1101) = 3$. In $[N,+]$, $2 + 3 = 5$. The class associated with 5 is [11111], which contains the correct answer of 1011101.

6. a) For $g,h \in S$, $f(g \circ h) = (g \circ h)(10) = g(h(10))$ and $f(g) + f(h) = g(10) + h(10)$. If $g(10) = 5$, $h(10) = 8$, and $g(8) = 17$, for example, then $g(h(10)) = g(8) = 17$ but $g(10) + h(10) = 5 + 8 = 13$.

b) Let $g,h \in S$. Then $g + h$ is a function from R to R so $g + h \in S$. Associativity holds because of associativity in $[R,+]$. The function z defined by $z(x) = 0$ for all $x \in R$ is a member of S and $g + z = z + g = g$ for all $g \in S$. For $g \in S$, define g' by $g'(x) = -g(x)$ for all $x \in R$. Then $g' \in S$ and $g + g' = g' + g = z$.

c) For $g,h \in S$, $f(g + h) = (g + h)(10) = g(10) + h(10) = f(g) + f(h)$. For $r \in R$, define a constant function $h(x) = r$ for all $x \in R$. Then $h \in S$ and $f(h) = h(10) = r$.

7. a) Closure holds, 1 is an identity, each element is self-inverse.

b) Consider four cases:
    i) α even, β even: then $f(\alpha) = 1$, $f(\beta) = 1$, and $\alpha \circ \beta$ is even. Thus $f(\alpha \circ \beta) = 1 = f(\alpha) \cdot f(\beta)$.
    ii) α odd, β even: then $f(\alpha) = -1$, $f(\beta) = 1$, and $\alpha \circ \beta$ is odd. Thus $f(\alpha \circ \beta) = -1 = f(\alpha) \cdot f(\beta)$.
    iii) α even, β odd: similar to (ii).
    iv) α odd, β odd: then $f(\alpha) = -1$, $f(\beta) = -1$, and $\alpha \circ \beta$ is even. Thus $f(\alpha \circ \beta) = 1 = f(\alpha) \cdot f(\beta)$.

8. Define a function $f: \mathcal{P}(S) \to \mathcal{P}(S)$ by $f(X) = S - X$. f is onto: for $X \in \mathcal{P}(S)$, $S - X \in \mathcal{P}(S)$ and $f(S - X) = S - (S - X) = S \cap (S \cap X')' = S \cap (S' \cup X) = (S \cap S') \cup (S \cap X) = \phi \cup X = X$. f is one-to-one: if $f(X) = f(Y)$, then $S - X = S - Y$ and $X = Y$. f is a homomorphism: for $X,Y \in \mathcal{P}(S)$, $f(X \cap Y) = S - (X \cap Y) = S \cap (X \cap Y)' = S \cap (X' \cup Y') = (S \cap X') \cup (S \cap Y') = (S - X) \cup (S - Y) = f(X) \cup f(Y)$.

*9. $f(x \cdot y) = i = i \cdot i = f(x) \cdot f(y)$

10. a) $f(x \cdot y) = (x \cdot y)^n = x^n \cdot y^n = f(x) \cdot f(y)$
    b) $f(x \cdot y) = (x \cdot y)^{-1} = y^{-1} \cdot x^{-1} = x^{-1} \cdot y^{-1} = f(x) \cdot f(y)$
    f is onto: let $g \in G$. Then $g^{-1} \in G$ and $f(g^{-1}) = (g^{-1})^{-1} = g$.
    f is one-to-one: if $f(x) = f(y)$ then $x^{-1} = y^{-1}$; because
    inverses are unique, $(x^{-1})^{-1} = (y^{-1})^{-1}$ or $x = y$.

11. a) Let $g \in Q$. The equation $nx = g$ has $x = g/n \in Q$ as a solution.
    b) $f(G)$ is a group by Theorem 7.67. Let $f(g) \in f(G)$ and consider the equation $x^n = f(g)$. Because $g \in G$ and G is divisible, $x^n = g$ has a solution g_1 in G, so that $g_1^n = g$. Then $f(g_1) \in f(G)$ and $(f(g_1))^n = f(g_1) \cdot f(g_1) \cdots f(g_1)$
    $= f(g_1^n) = f(g)$.

*12. Because f is an onto function, $U \neq \phi$. To show closure, let $x, y \in U$. Then $f(x \cdot y) = f(x) + f(y) = t + t = t$, so $x \cdot y \in U$.

*13. a) Let x be an idempotent element in $[S, \cdot]$. Then $x^2 = x$. Also, $(f(x))^2 = f(x) + f(x) = f(x \cdot x) = f(x)$, and $f(x)$ is idempotent in $[T, +]$.
    b) For any $f(x) \in T$, $f(x) + f(0) = f(x \cdot 0) = f(0)$ and $f(0) + f(x) = f(0 \cdot x) = f(0)$, so $f(0)$ is a zero in $[T, +]$.

14. a) [Aut(S), o] is closed because composition of isomorphisms is an isomorphism (Practice 7.73). Associativity always holds for function composition. The identity function i_S is an automorphism on S. Finally, if f is an automorphism on S, so is f^{-1}.
    b)
    i: $0 \to 0$ f: $0 \to 0$
    $1 \to 1$ $1 \to 3$
    $2 \to 2$ $2 \to 2$
    $3 \to 3$ $3 \to 1$

o	i	f
i	i	f
f	f	i

15. a) For $f, g \in \text{Hom}(G)$, $f + g: G \to G$, and $(f + g)(x + y)$
    $= f(x + y) + g(x + y) = f(x) + f(y) + g(x) + g(y)$
    (because f and g are homomorphisms) $= f(x) + g(x) + f(y) + g(y)$ (because $[G, +]$ is commutative) $= (f + g)(x) + (f + y)(y)$; therefore $(f + g) \in \text{Hom}(G)$. To show associativity, let $f, g, h \in \text{Hom}(G)$; then $(f + (g + h))(x)$
    $= f(x) + (g + h)(x) = f(x) + (g(x) + h(x))$
    $= (f(x) + g(x)) + h(x)$ (because $[G, +]$ is associative)

$= (f + g)(x) + h(x) = ((f + g) + h)(x)$. To show
commutativity, let $f,g \in \text{Hom}(G)$; then $(f + g)(x)$
$= f(x) + g(x) = g(x) + f(x) = (g + f)(x)$.

b) $i(x + y) = 0_G = 0_G + 0_G = i(x) + i(y)$, so $i \in \text{Hom}(G)$. For
$f \in \text{Hom}(G)$, $(f + i)(x) = f(x) + i(x) = f(x) + 0_G = f(x)$
and by commutativity, $(i + f)(x) = f(x)$. Therefore i is
an identity of $[\text{Hom}(G),+]$.

c) $(-f)(x + y) = -f(x + y) = -(f(x) + f(y))$
$= (-f(y)) + (-f(x))$ (by Theorem 7.30) $= (-f(x)) + (-f(y))$
$= (-f)(x) + (-f)(y)$, so $-f \in \text{Hom}(G)$. Also, $(f + (-f))(x)$
$= f(x) + (-f(x)) = 0_G$ and $((-f) + f)(x) = 0_G$, so
$f + (-f) = (-f) + f = i$ and $-f$ is the inverse of f in
$[\text{Hom}(G),+]$.

16. Let f be an isomorphism from G to H. Then f is a bijection and f^{-1} exists. f^{-1} is a homomorphism, $f^{-1}: H \longrightarrow G$, and $f^{-1} \circ f = i_G$, $f \circ f^{-1} = i_H$. Now let f be a homomorphism from G to H. If there exists a function $g: H \longrightarrow G$ such that $g \circ f = i_G$ and $f \circ g = i_H$, then f is a bijection (see Exercise 20, Section 3.2), hence an isomorphism.

17. Let i_G and i_H denote the identity elements of G and H, respectively. Let f be an isomorphism, $f: G \longrightarrow H$. Then $f(i_G) = i_H$ by Theorem 7.67 and since f is one-to-one, i_G is the only such element. Now let f be a homomorphism from G onto H; then $f(i_G) = i_H$. Suppose i_G is the only such element, and let $f(g_1) = f(g_2)$ for $g_1, g_2 \in G$. Then $f(g_1 \cdot g_2^{-1}) = f(g_1) \cdot f(g_2^{-1}) = f(g_1) \cdot (f(g_2))^{-1} = f(g_1) \cdot (f(g_1))^{-1} = i_H$. Therefore $g_1 \cdot g_2^{-1} = i_G$ and $g_1 = i_G \cdot g_2 = g_2$. Thus f is one-to-one; f is already an onto homomorphism, so it is an isomorphism.

*18. a) Let $f: G \times H \longrightarrow H \times G$ be defined by $f((x,y)) = (y,x)$.
Then f is one-to-one and onto, and $f((x_1,y_1) \cdot (x_2,y_2))$
$= f((x_1 \cdot x_2, y_1 + y_2)) = (y_1 + y_2, x_1 \cdot x_2) = (y_1, x_1) \cdot (y_2, x_2)$
$= f((x_1,y_1)) \cdot f((x_2,y_2))$

b) $f((x_1,y_1) \cdot (x_2,y_2)) = f((x_1 \cdot x_2, y_1 + y_2)) = x_1 \cdot x_2$
$= f((x_1,y_1)) \cdot f((x_2,y_2))$

*19. Let $x,y \in G$. Then $f(x \cdot y) = (x \cdot y)^{-1} = y^{-1} \cdot x^{-1} = f(y) \cdot f(x)$
$= f(y \cdot x)$. Therefore $x \cdot y = y \cdot x$ because f is one-to-one.

20. For each $s \in S$, define a mapping $\alpha_s: S \to S$ by $\alpha_s(x) = s \cdot x$. Let $T = \{\alpha_s \mid s \in S\}$. Then $[T, \circ]$ is a semigroup because $T \neq \phi$ (since $S \neq \phi$) and $[T, \circ]$ is closed. To prove this, let $\alpha_s, \alpha_r \in T$. Then $(\alpha_s \circ \alpha_r)(x) = \alpha_s(\alpha_r(x)) = \alpha_s(r \cdot x) = s \cdot (r \cdot x) = (s \cdot r) \cdot x = \alpha_{s \cdot r}(x)$, so $\alpha_s \circ \alpha_r = \alpha_{s \cdot r} \in T$. Therefore T is a transformation semigroup.

Let $f: S \to T$ be given by $f(s) = \alpha_s$ for $s \in S$. Then f is clearly onto. To show that f is a homomorphism, let $s, r \in S$. Then $f(s \cdot r) = \alpha_{s \cdot r} = \alpha_s \circ \alpha_r = f(s) \circ f(r)$.

21. To show that f is onto, let $y \in G$. Then $g^{-1} \cdot y \cdot g \in G$ and $f(g^{-1} \cdot y \cdot g) = g \cdot (g^{-1} \cdot y \cdot g) \cdot g^{-1} = (g \cdot g^{-1}) \cdot y \cdot (g \cdot g^{-1}) = y$. To show that f is one-to-one, $f(x) = f(y) \to g \cdot x \cdot g^{-1} = g \cdot y \cdot g^{-1}$
$\to x = y$ by cancellation. To show that f is a homomorphism, $f(x \cdot y) = g \cdot (x \cdot y) \cdot g^{-1} = g \cdot x \cdot (g^{-1} \cdot g) \cdot y \cdot g^{-1} = (g \cdot x \cdot g^{-1}) \cdot (g \cdot y \cdot g^{-1})$
$= f(x) \cdot f(y)$.

Section 7.4

*1. a) $f(a_2 x^2 + a_1 x + a_0 + b_2 x^2 + b_1 x + b_0)$
$= f((a_2 + b_2)x^2 + (a_1 + b_1)x + (a_0 + b_0)) = a_2 + b_2$
$= f(a_2 x^2 + a_1 x + a_2) + f(b_2 x^2 + b_1 x + b_0)$
Clearly f is onto.

b) (i) and (iii)

2. *a) $f(Z_{18}) = \{0, 4, 8\}$
$Z_{18}/f = \{[0], [1], [2]\}$ where
$[0] = \{0, 3, 6, 9, 12, 15\}$
$[1] = \{1, 4, 7, 10, 13, 16\}$
$[2] = \{2, 5, 8, 11, 14, 17\}$

The isomorphism is
$[0] \to 0 \qquad [1] \to 4 \qquad [2] \to 8$

b) $f(Z) = \{0, 1, 2, 3, 4, 5, 6\} = Z_7$
$Z/f = \{[0], [1], [2], [3], [4], [5], [6]\}$ where
$[0] = \{0, 7, 14, \ldots, -7, -14, \ldots\}$
$[1] = \{1, 8, 15, \ldots, -6, -13, \ldots\}$
$[2] = \{2, 9, 16, \ldots, -5, -12, \ldots\}$
$[3] = \{3, 10, 17, \ldots, -4, -11, \ldots\}$
$[4] = \{4, 11, 18, \ldots, -3, -10, \ldots\}$
$[5] = \{5, 12, 19, \ldots, -2, -9, \ldots\}$
$[6] = \{6, 13, 20, \ldots, -1, -8, \ldots\}$

The isomorphism is
$$[0] \to 0 \qquad [4] \to 4$$
$$[1] \to 1 \qquad [5] \to 5$$
$$[2] \to 2 \qquad [6] \to 6$$
$$[3] \to 3$$

c) $f(Z_{12}) = \{0,6,12\}$
$Z_{12}/f = \{[0],[1],[2]\}$ where
$$[0] = \{0,3,6,9\}$$
$$[1] = \{1,4,7,10\}$$
$$[2] = \{2,5,8,11\}$$

The isomorphism is
$$[0] \to 0 \qquad [1] \to 6 \qquad [2] \to 12$$

d) $f(M_2^0(Q)) = \{(0,q) \mid q \in Q\}$

$M_2^0(Q)/f$ consists of classes each containing a single matrix of the form $\begin{bmatrix} 1 & q \\ 0 & 1 \end{bmatrix}$.

The isomorphism maps the class containing $\begin{bmatrix} 1 & q \\ 0 & 1 \end{bmatrix}$ to $(0,q)$.

3. *a) $g([1] * [2]) = g([1 +_{18} 2]) = g([3]) = g([0]) = 0$
$g([1]) +_{12} g([2]) = 4 +_{12} 8 = 0$

b) $g([3] * [3]) = g([3 + 3]) = g([6]) = 6$
$g([3]) +_7 g([3]) = 3 +_7 3 = 6$

c) $g([2] * [2]) = g([2 +_{12} 2]) = g([4]) = g([1]) = 6$
$g([2]) +_{18} g([2]) = 12 +_{18} 12 = 6$

d) $g\left(\begin{bmatrix} 1 & 5 \\ 0 & 1 \end{bmatrix} * \begin{bmatrix} 1 & -4 \\ 0 & 1 \end{bmatrix}\right) = g\left(\begin{bmatrix} 1 & 5 \\ 0 & 1 \end{bmatrix} \cdot \begin{bmatrix} 1 & -4 \\ 0 & 1 \end{bmatrix}\right)$
$= g\left(\begin{bmatrix} 1 & 1 \\ 0 & 1 \end{bmatrix}\right) = (0,1)$

$g\left(\begin{bmatrix} 1 & 5 \\ 0 & 1 \end{bmatrix}\right) + g\left(\begin{bmatrix} 1 & -4 \\ 0 & 1 \end{bmatrix}\right) = (0,5) + (0,-4) = (0,1)$

4. a) Even: $\alpha_1 = i = (1,2)\circ(1,2)$
$\alpha_5 = (1,2,3) = (1,3)\circ(1,2)$
$\alpha_6 = (1,3,2) = (1,2)\circ(1,3)$

Odd: $\alpha_2 = (1,2)$
$\alpha_3 = (1,3)$
$\alpha_4 = (2,3)$

b) $S_3/f = \{[\alpha_1],[\alpha_2]\}$ where $[\alpha_1] = \{\alpha_1,\alpha_5,\alpha_6\}$ and

$[\alpha_2] = \{\alpha_2, \alpha_3, \alpha_4\}$

c) The isomorphism is
$[\alpha_1] \to 1 \qquad [\alpha_2] \to -1$

d) $g^{-1}(1) = [\alpha_1]$, $g^{-1}(-1) = [\alpha_2]$; $[\alpha_1]*[\alpha_2] = [\alpha_1 \circ \alpha_2]$
$= [i \circ (1,2)] = [(1,2)] = [\alpha_2]$ and $g([\alpha_2]) = -1$.

5. f: $0 \to 0 \qquad 4 \to 0 \qquad$ or $f(x) = 1 \cdot 4x$ (homomorphism
$1 \to 1 \qquad 5 \to 1 \qquad$ from Z_8 onto Z_4)
$2 \to 2 \qquad 6 \to 2$
$3 \to 3 \qquad 7 \to 3$

6. $i_S \in [i_S]$ so $[i_S] \neq \phi$. Let $x, y \in [i_S]$. Then $f(x) = f(y) = i_T$.
Also, $f(x \cdot y^{-1}) = f(x) + f(y^{-1}) = f(x) + (-f(y)) = i_T + (-i_T)$
$= i_T$. Therefore $x \cdot y^{-1} \in [i_S]$ and $[i_S]$ is a subgroup by Exercise 18a of Section 7.2.

Section 7.5

1. a) $0 + 8Z = \{0, \pm 8, \pm 16, \ldots\}$ b) $(3 + 8Z) + (6 + 8Z)$
$1 + 8Z = \{1, 9, 17, \ldots, -7, -15, \ldots\}$ $= 1 + 8Z$
$2 + 8Z = \{2, 10, 18, \ldots, -6, -14, \ldots\}$
$3 + 8Z = \{3, 11, 19, \ldots, -5, -13, \ldots\}$
$4 + 8Z = \{4, 12, 20, \ldots, -4, -12, \ldots\}$
$5 + 8Z = \{5, 13, 21, \ldots, -3, -11, \ldots\}$
$6 + 8Z = \{6, 14, 22, \ldots, -2, -10, \ldots\}$
$7 + 8Z = \{7, 15, 23, \ldots, -1, -9, \ldots\}$

2. a) i, (1,2,3,4), (1,2,4,3), (1,3,2,4), (1,3,4,2), (1,4,2,3), (1,4,3,2), (1,2), (1,3), (1,4), (2,3), (2,4), (3,4), (1,2,3), (1,3,2), (1,2,4), (1,4,2), (1,3,4), (1,4,3), (2,3,4), (2,4,3), (1,2)∘(3,4), (1,3)∘(2,4), (1,4)∘(2,3)

b) iS = S
(1,2,3,4)S = {(1,2,3,4)∘i, (1,2,3,4)∘(1,2)∘(3,4),
                 (1,2,3,4)∘(1,4)∘(2,3), (1,2,3,4)∘(1,3)∘(2,4)}
           = {(1,2,3,4), (1,3), (2,4), (1,4,3,2)}
(1,2,4,3)S = {(1,2,4,3), (1,4), (1,3,4,2), (2,3)}
(1,3,2,4)S = {(1,3,2,4), (1,4,2,3), (3,4), (1,2)}
(1,2,3)S = {(1,2,3), (1,3,4), (1,4,2), (2,4,3)}
(1,2,4)S = {(1,2,4), (1,4,3), (2,3,4), (1,3,2)}

*3. a) $0 + S = \{0,4,8\}$
$1 + S = \{1,5,9\}$
$2 + S = \{2,6,10\}$
$3 + S = \{3,7,11\}$
b) $(2 + S) + (3 + S) = 1 + S$

4. a) $(0,0) + S = \{(0,0),(1,2),(2,0),(0,2),(1,0),(2,2)\} = S$
$(1,1) + S = \{(1,1),(2,3),(0,1),(1,3),(2,1),(0,3)\}$
b) $(1,1) + S + (1,1) + S = (0,0) + S$

*5. a) $1S = \{1,2,4\}$
$3S = \{3,6,5\}$
b) $3S \cdot 4S = 3S$

6. a) $K = 8Z$
$G/K = Z/8Z = \{0+8Z, 1+8Z, 2+8Z, 3+8Z, 4+8Z, 5+8Z, 6+8Z, 7+8Z\}$
b) $K = \{0,4,8\}$
$G/K = Z_{12}/K = \{0 + K, 1 + K, 2 + K, 3 + K\}$
where $0 + K = \{0,4,8\}$, $1 + K = \{1,5,9\}$, $2 + K = \{2,6,10\}$
$3 + K = \{3,7,11\}$
c) $K = \{(x,1/x) \mid x \in R^*\}$
G/K is the collection of all distinct sets of the form
$(a,b) \cdot K = \{(ax,b/x) \mid x \in R^*\}$ where $a,b \in R^*$.
d) K is the set of all polynomials in x over R with constant term equal to 0. A typical member of G/K is the set of all polynomials in x over R with constant term c.
e) $K = \left\{ \begin{bmatrix} a & b \\ c & a \end{bmatrix} \middle| a,b,c \in Q \right\}$. A typical member of G/K is
$\begin{bmatrix} x & y \\ u & v \end{bmatrix} + K = \left\{ \begin{bmatrix} x+a & y+b \\ u+c & v+a \end{bmatrix} \middle| a,b,c \in Q \right\}$ where
$x,y,u,v \in Q$.

7. $K = \{x \mid f(x) = x \cdot_k 1 = 0\} = \{x \mid x \text{ is a multiple of } k\} = kZ$

8. a) For $z \in Z$, $f((z,0)) = z + 0 = z$, so f is onto. Also
$f((x_1,y_1) + (x_2,y_2)) = f((x_1 + x_2, y_1 + y_2)) = x_1 + x_2 + y_1 + y_2 = x_1 + y_1 + x_2 + y_2 = f((x_1,y_1)) + f((x_2,y_2))$,
so f is a homomorphism.
b) $K = \{(z,-z) \mid z \in Z\}$.
c) $(4,1), (1,4), (5,0)$
d) $g: (Z \times Z)/K \longrightarrow Z$ defined by $g((x,y) + K) = x + y$

*9. a) $f(Q \times Q \times Q) = \{(y,y) | y \in Q\}$
   $f((x_1,y_1,z_1) + (x_2,y_2,z_2)) = f((x_1 + x_2, y_1 + y_2, z_1 + z_2))$
   $= (y_1 + y_2, y_1 + y_2) = (y_1,y_1) + (y_2,y_2)$
   $= f((x_1,y_1,z_1)) + f((x_2,y_2,z_2))$
   b) $K = \{(x,0,z) | x,z \in Q\}$
   c) $(4,4,4)$ and $(5,4,3)$
   d) Yes
   e) $g: (Q \times Q \times Q)/K \longrightarrow f(Q \times Q \times Q)$ defined by
      $g((x,y,z) + K) = (y,y)$

10. a) $f(Q*)$ is the set of all positive rational numbers with rational square roots.
      $f(xy) = (xy)^2 = x^2 y^2 = f(x)f(y)$
    b) $K = \{1,-1\}$
    c) $(3K)(7K) = 21K = \{21,-21\}$
    d) $g: Q*/K \longrightarrow f(Q*)$ defined by $g(xK) = x^2$

11. a) For $x \in G$, $f(x, i_H) = x$ so f is onto.
      $f((x_1,y_1) \cdot (x_2,y_2)) = f(x_1 \cdot x_2, y_1 \cdot y_2) = x_1 \cdot x_2$
      $= f(x_1,y_1) \cdot f(x_2,y_2)$, so f is a homomorphism
    b) $K = \{(i_G, h) | h \in H\}$
    c) $g: (G \times H)/K \rightarrow G$ defined by $g((x,y)K) = x$

12. a) Closure: $f + g \in F$. Associativity: $((f + g) + h)(x)$
      $= (f + g)(x) + h(x) = (f(x) + g(x)) + h(x)$
      $= f(x) + (g(x) + h(x)) = f(x) + (g + h)(x)$
      $= (f + (g + h))(x)$. The zero function $0(x) = 0$ for all x
      belongs to F; for $f \in F$, $-f \in F$ where $(-f)(x) = -f(x)$
      for all real x.
    b) For any $r \in R$, the function f_r, defined by $f_r(x) = r$ for
      all real x, is a member of F and $\alpha(f_r) = f_r(a) = r$,
      therefore α is onto. Also, for $f,g \in F$, $\alpha(f + g)$
      $= (f + g)(a) = f(a) + g(a) = \alpha(f) + \alpha(g)$.
    c) $K = \{f | f \in F$ and $\alpha f = f(a) = 0\}$, so K is the set of all
      functions whose value at a is 0.
    d) For $h \in g + K$, $h = g + f$ where $f \in K$. Thus $h(a)$
      $= (g + f)(a) = g(a) + f(a) = g(a) + 0 = g(a)$.
    e) $\theta: F/K \longrightarrow R$ defined by $\theta(g + K) = g(a)$

*13. (This is Exercise 17 of Section 7.3).
Let i_G and i_H denote the identity elements of G and H. Let f be an isomorphism, f: G $\to$ H. Then $f(i_G) = i_H$ by Theorem 7.67, and because f is one-to-one, i_G is the only element mapping to i_H. Thus $K = \{i_G\}$.

Now assume that $K = \{i_G\}$; we need to show that the homomorphism f is one-to-one. Let g_1 and g_2 be elements of G with $f(g_1) = f(g_2)$. Then $f(g_1 \cdot g_2^{-1}) = f(g_1) \cdot f(g_2^{-1})$
$= f(g_1) \cdot (f(g_2))^{-1} = f(g_1) \cdot (f(g_1))^{-1} = i_H$. Therefore $g_1 \cdot g_2^{-1} = i_G$ and $g_1 = i_G \cdot g_2 = g_2$.

*14. Let $x \in A$ and $g \in G$; we want to show that $g^{-1}xg \in A$. But since $x \in A$, $g^{-1}xg = g^{-1}gx = x \in A$. A is normal by Theorem 7.113.

15. $|S_n| = |A_n| \cdot m$ where m is the number of cosets, i.e., $m = [S_n : A_n] = |S_n/A_n|$. $|S_n| = n!$ and $|A_n| = n!/2$, so $m = 2$.

16. Let $n = kq$. Because $[Z_n, +_n]$ is commutative, the set $(k) = \{k, k +_n k = 2k, 3k, \ldots (q-1)k, qk = n = 0\}$ is a normal subgroup of order q. The distinct cosets in $Z_n/(k)$ are $0 + (k)$, $1 + (k), 2 + (k), \ldots k - 1 + (k)$, so that the order of $Z_n/(k)$ is k.

17. a) Let S be a normal subgroup of a commutative group G. Then in G/S, $xS \cdot yS = (x \cdot y)S = (y \cdot x)S = yS \cdot xS$.
    b) Let S be a normal subgroup of a divisible group G, and let n be a positive integer. Let $gS \in G/S$. Then $g \in G$ and there exists $h \in G$ such that $h^n = g$, so $(hS)^n = h^n S = gS$. Thus hS is a solution in G/S to the equation $x^n = gS$.

*18. Let S be normal in G and let $x \in G$, $s \in S$. Because S is normal, $sx \in Sx = xS$ so that $sx = xs'$ for some $s' \in S$. Then $x^{-1}sx = x^{-1}xs' = s' \in S$.

19. a) G is closed under ·
    Associativity holds:
    $((a,b) \cdot (c,d)) \cdot (e,f) = (ac, ad + b) \cdot (e,f) = ((ac)e, (ac)f + ad + b)$
    $(a,b) \cdot ((c,d) \cdot (e,f)) = (a,b) \cdot (ce, cf + d) = (a(ce), a(cf + d) + b)$
    Identity:
    $(a,b) \cdot (1,0) = (a \cdot 1, a \cdot 0 + b) = (a,b)$
    $(1,0) \cdot (a,b) = (1 \cdot a, 1 \cdot b + 0) = (a,b)$

Inverses:

$(a,b)(\frac{1}{a},\frac{-b}{a}) = (a\cdot\frac{1}{a}, a(\frac{-b}{a}) + b) = (1,0)$

$(\frac{1}{a},\frac{-b}{a})(a,b) = (\frac{1}{a}\cdot a, \frac{1}{a}\cdot b - \frac{b}{a}) = (1,0)$

b) S is closed under $\cdot$, $(1,0) \in S$, and for $(a,0) \in S$,
$(a,0)^{-1} = (\frac{1}{a},0) \in S$
T is closed under $\cdot$, $(1,0) \in T$, and for $(1,b) \in T$,
$(1,b)^{-1} = (1,-b) \in T$

c) For $(a,b) \in G$, $(c,0) \in S$, $(a,b)^{-1}\cdot(c,0)\cdot(a,b) =$
$(\frac{1}{a},\frac{-b}{a})\cdot(c,0)\cdot(a,b) = (\frac{c}{a},\frac{-b}{a})\cdot(a,b) = (c, \frac{cb}{a} - \frac{b}{a})$, which
belongs to S only for $c = 1$. Hence S is not normal, by Exercise 18.

For $(a,b) \in G$, $(1,c) \in T$, $(a,b)^{-1}\cdot(1,c)\cdot(a,b) =$
$(\frac{1}{a},\frac{-b}{a})\cdot(1,c)\cdot(a,b) = (\frac{1}{a},\frac{c}{a} - \frac{b}{a})\cdot(a,b) = (1,\frac{b}{a} + \frac{c}{a} - \frac{b}{a}) =$
$(1,\frac{c}{a}) \in T$. Hence T is normal, by Theorem 7.113.

20. a) $S \cap T$ is a subgroup (Exercise 9a, Section 7.2). Let $x \in G$ and $a \in S \cap T$. Because S is normal, $x^{-1}ax \in S$ and because T is normal $x^{-1}ax \in T$ (Exercise 14 above). Therefore $x^{-1}ax \in S \cap T$ and $S \cap T$ is normal by Theorem 7.113.

b) Let $(x_1,y_1),(x_2,y_2) \in S_1 \times S_2$. Then $(x_1,y_1)\cdot(x_2,y_2)^{-1}$
$= (x_1,y_1)\cdot(x_2^{-1},y_2^{-1}) = (x_1\cdot x_2^{-1}, y_1\cdot y_2^{-1}) \in S_1 \times S_2$;
therefore $S_1 \times S_2$ is a subgroup by Exercise 18a of Section 7.2. To show that $S_1 \times S_2$ is normal in $G_1 \times G_2$, let $(x_1,y_1) \in S_1 \times S_2$ and let $(x,y) \in G_1 \times G_2$. S_1 is normal in G_1, so $x^{-1}x_1x \in S_1$; S_2 is normal in G_2 so $y^{-1}y_1y \in S_2$ (Exercise 18 above). Then
$(x,y)^{-1}(x_1,y_1)(x,y) = (x^{-1},y^{-1})(x_1,y_1)(x,y)$
$= (x^{-1}x_1x, y^{-1}y_1y) \in S_1 \times S_2$, and $S_1 \times S_2$ is normal by Theorem 7.113.

21. a) $g^{-1}Hg$ is a nonempty subset of G. If $g^{-1}h_1g$ and $g^{-1}h_2g$ are members of $g^{-1}Hg$, then $(g^{-1}h_1g)(g^{-1}h_2g)^{-1}$
$= (g^{-1}h_1g)(g^{-1}h_2^{-1}g) = g^{-1}h_1h_2^{-1}g \in g^{-1}Hg$. Thus $g^{-1}Hg$ is a subgroup by Exercise 18a of Section 7.2.

b) Let $x \in G$, $h \in H$. By part(a) $x^{-1}Hx$ is a subgroup of G. Let $f: x^{-1}Hx \to H$ be defined by $f(x^{-1}hx) = h$. Then f is onto and if $f(x^{-1}h_1x) = f(x^{-1}h_2x)$, then $h_1 = h_2$ and

$x^{-1}h_1x = x^{-1}h_2x$, so f is one-to-one. Therefore $|x^{-1}Hx| = |H|$, so $x^{-1}Hx = H$, or $Hx = xH$ and H is normal.

22. $f(S)$ is a subgroup of $f(G)$ by Theorem 7.67. To show that it is normal, let $f(x) \in f(G)$ and $f(s) \in f(S)$. Then $x \in G$, $s \in S$, and because S is normal in G, $x^{-1}sx \in S$ (Exercise 18 above). Then $(f(x))^{-1} \cdot f(s) \cdot f(x) = f(x^{-1}) \cdot f(s) \cdot f(x) = f(x^{-1}sx) \in f(S)$, and $f(S)$ is normal in $f(G)$ by Theorem 7.113.

*23. For any $x \in G$, $xT = Tx$ because T is normal in G; thus $xT = Tx$ for any $x \in S$ and T is normal in S. To show that S/T is a subgroup of G/T, note that $S/T \subseteq G/T$ and let $xT, yT \in S/T$. Then $x,y \in S$ so $xy^{-1} \in S$ and $xT(yT)^{-1}$ $= xT(y^{-1}T) = xy^{-1}T \in S/T$. To show that S/T is normal in G/T, let $sT \in S/T$ and $xT \in G/T$. Then, because S is normal, $x^{-1}sx \in S$ (Exercise 18 above) and thus $(xT)^{-1}(sT)(xT)$ $= (x^{-1}T)(sT)(xT) = (x^{-1}sx)T \in S/T$; apply Theorem 7.113.

24. a) Assume that $xT = yT$ and let $xs \in xS$. Let $t_1 \in T$. Then $xt_1 \in xT = yT$, so $xt_1 = yt_2$ for some $t_2 \in T$. Then $xs = xt_1t_1^{-1}s = yt_2t_1^{-1}s \in yS$, and $xS \subseteq yS$. Similarly, $yS \subseteq xS$.

b) Let $x_1T, x_2T \in G/T$. Then $f(x_1Tx_2T) = f(x_1x_2T) = x_1x_2S$ $= x_1Sx_2S = f(x_1T)f(x_2T)$.

c) $K = \{xT| \, x \in G$ and $f(xT) = xS = S\}$. But $xS = S$ if and only if $x \in S$. Hence $K = \{xT| \, x \in S\} = S/T$.

d) Because the homomorphism f is an onto function with S/T as its kernel, the result follows from Theorem 7.116.

25. a) f_g is onto: For $x \in G$, $g^{-1}xg \in G$ and $f_g(g^{-1}xg)$ $= gg^{-1}xgg^{-1} = x$. f_g is one-to-one: if $f(x) = f(y)$, then $gxg^{-1} = gyg^{-1}$ and $x = y$ by cancellation. f_g is a homomorphism: $f_g(xy) = gxyg^{-1} = gxg^{-1}gyg^{-1} = f_g(x)f_g(y)$.

b) Composition is associative. Closure: let f_g and f_h belong to F. Then for $x \in G$, $(f_g \circ f_h)(x) = f_g(hxh^{-1})$ $= ghxh^{-1}g^{-1} = (gh)x(gh)^{-1} = f_{gh}(x)$, and $f_{gh} \in F$. Identity: let i be the identity of G. Then $f_i \in F$ and $f_g \circ f_i$ $= f_{g \cdot i} = f_g$ and $f_i \circ f_g = f_g$. Inverses: for $f_g \in F$, $(f_g)^{-1} = f_{g^{-1}}$ because $f_g \circ f_{g^{-1}} = f_i = f_{g^{-1}} \circ f_g$.

c) Define a function θ: $G \longrightarrow F$ by $\theta(g) = f_g$. Then θ is a

133

homomorphism because $\theta(gh) = f_{gh} = f_g \circ f_h = \theta(g) \circ \theta(h)$. Also $\theta(G) = F$. The kernel K of θ is $\{g \in G| \theta(g) = f_g = f_i\}$. The condition $f_g = f_i$ says that for all $x \in G$, $f_g(x) = gxg^{-1} = x$. But $gxg^{-1} = x$ if and only if $gx = xg$, or $g \in A$. Thus $K = A$, and the result follows from Theorem 7.116.

26. There are two left cosets of S in G; one is S and the other must be $G - S$. Similarly the two right cosets of S in G are S and $G - S$, so left and right cosets coincide and S is normal.

27. By Exercise 26, S is a normal subgroup of G. Therefore the factor group G/S exists, and $|G/S| = 2$; the two cosets are S and $G - S$. By Theorem 7.74, $G/S \simeq G_2$ where the table for G_2 is

·	1	a
1	1	a
a	a	1

    Let g be the isomorphism from G/S to G_2. Then $g(S) = 1$ (because S is the identity in G/S), and $g(G - S) = a$. The mapping $f: G \to G/S$ given by
    $$f(x) = \begin{cases} S & \text{if } x \in S \\ G - S & \text{if } x \notin S \end{cases}$$
    is a homomorphism (Practice 7.115). Thus $g \circ f$ is a homomorphism from G to G_2 given by
    $$(g \circ f)(x) = g(f(x)) = \begin{cases} g(S) = 1 & \text{if } x \in S \\ g(G - S) = a & \text{if } x \notin S \end{cases}$$
    Then for $x \in G$, $(g \circ f)(x^2) = (g \circ f)(x) \cdot (g \circ f)(x) = ((g \circ f)(x))^2 = 1$, whether $x \in S$ or $x \notin S$. Therefore $x^2 \in S$.

*28. $[G:T] = \frac{|G|}{|T|} = \frac{|G|}{|S|} \cdot \frac{|S|}{|T|} = [G:S] \cdot [S:T]$

29. Let $f: L \to R$ be given by $f(xS) = Sx^{-1}$. Then f is well-defined because if $xS = yS$, then $y = sx$ for some $s \in S$, and $f(yS) = Sy^{-1} = S(s^{-1}x^{-1}) = Sx^{-1} = f(xS)$. To show that f is one-to-one, let $f(x_1 S) = f(x_2 S)$. Then $Sx_1^{-1} = Sx_2^{-1}$ so $x_2^{-1} = sx_1^{-1}$ for some $s \in S$, or $x_2 = x_1 s^{-1}$ and $x_2 S = x_1 s^{-1} S = x_1 S$. Finally f is onto because for $Sy \in R$, $y^{-1} S \in L$ and $f(y^{-1} S) = Sy$.

CHAPTER 8

Section 8.1

*1. a) s = 12; ALLGAULISDIVIDED
   b) There are no occurrences of the letter E.
   c) IBM

2. a) $M^{-1} = \begin{bmatrix} 5 & 24 \\ 19 & 3 \end{bmatrix}$ (remember that this is arithmetic in Z_{26}); decoded message is ATTACK.
   b) ATMDCK

3. a) We must have $x \cdot 8 \, 11 = 1$, $0 < x < 8$, so $x = 3$.
   b) The code for 3 is $3^3 \cdot_{15} 1 = 27 \cdot_{15} 1 = 12$
   c) To decode, compute $12^{11} \cdot_{15} 1$. Doing successive reductions modulo 15, $12^{11} \to (12^2)^5 \cdot 12 \to 9^5 \cdot 12 \to 81 \cdot 81 \cdot 9 \cdot 12 \to 6 \cdot 6 \cdot 9 \cdot 12 \to 6 \cdot 9 \cdot 12 \to 9 \cdot 12 \to 3$

4. *a) Yes
   *b) 152478; 152748 (error detected); 125748 (no error detected)
   c) Points to include: The individual digits in the string must be selected and the required arithemetic done; the last digit of the integer answer must be isolated and compared with the last string digit.

5. a) $A(n,n) = 2$ (To have minimum distance n, the n-tuple code words must differ in n components; such words only occur in pairs.)
   b) $A(n,1) = 2^n$ (The set of all binary n-tuples has minimum distance 1.)
   c) $A(n,2) = 2^{n-1}$ (The set of all possibilities for the first n − 1 positions gives minimum distance ≥ 1; to get distance ≥ 2, the n<u>th</u> component is fixed once the first code word is chosen.)

6. Let $X = (x_1, \ldots, x_n)$ and $Y = (y_1, \ldots, y_n) \in Z_2^n$. The i<u>th</u> component of $(X +_2 Y)H$ is given by
   $$(x_1 + y_1)h_{1i} + (x_2 + y_2)h_{2i} + \cdots + (x_n + y_n)h_{ni}$$
   where additions and multiplications are all modulo 2. Because the distributive law holds and addition modulo 2 is commutative, this expression equals
   $$(x_1 h_{1i} + x_2 h_{2i} + \cdots + x_n h_{ni}) + (y_1 h_{1i} + y_2 h_{2i} + \cdots + y_n h_{ni})$$

which is the i<u>th</u> component of $X \cdot H +_2 Y \cdot H$.

7. a) H has no row of all 0's and no two rows alike, so the code is single-error correcting.
   b) $n = 6$, $r = 3$, $m = 6 - 3 = 3$. H can encode all of Z_2^3:
   000 → 000000
   001 → 001101
   010 → 010011
   100 → 100111
   110 → 110100
   101 → 101010
   011 → 011110
   111 → 111001

8. a) H has no row of all 0's and no two rows alike, so the code is single-error correcting.
   b) $n = 6$, $r = 3$, $m = 6 - 3 = 3$. H can encode all of Z_2^3:
   000 → 000000
   001 → 001011
   010 → 010110
   011 → 011101
   100 → 100101
   101 → 101110
   110 → 110011
   111 → 111000

*9. a) H has no row of all 0's and no two rows alike, so the code is single-error correcting.
   b) $n = 9$, $r = 4$, $m = 9 - 4 = 5$. H can encode all of Z_2^5:

   00000 → 000000000   11100 → 111001111
   00001 → 000011100   11010 → 110100011
   00010 → 000101001   11001 → 110010110
   00100 → 001000101   10110 → 101100001
   01000 → 010000111   10101 → 101010100
   10000 → 100001101   10011 → 100111000
   11000 → 110001010   01011 → 010110010
   10100 → 101001000   01101 → 011011110
   10010 → 100100100   00111 → 001110000
   10001 → 100010001   01110 → 011101011

136

```
01100 → 011000010        01111 → 011110111
01010 → 010101110        10111 → 101111101
01001 → 010011011        11011 → 110111111
00110 → 001101100        11101 → 111010011
00101 → 001011001        11110 → 111100110
00011 → 000110101        11111 → 111111010
```
    c) No; $m \neq 2^r - r - 1$

10. Points to include: m must be computed, and the members of Z_2^m generated. For each member of Z_2^m, a multiplication must be done with each of the r columns of H, counting the number of locations at which 1's match, and thus choosing the last r components one at a time. (Much nested looping should occur in actual code!)

*11. a) $n = 5$, $m = 3$, $r = 2$; $m \not\leq 2^r - r - 1$, so neither perfect nor single-error correcting.

    b) $n = 12$, $m = 7$, $r = 5$; $m < 2^r - r - 1$, so single-error correcting but not perfect.

    c) $n = 15$, $m = 11$, $r = 4$; $m = 2^r - r - 1$, so perfect

*12. a) 6 ($32 \leq 2^6 - 6 - 1$)

    b) 6 ($36 \leq 2^6 - 6 - 1$)

    c) 7 ($60 \leq 2^7 - 7 - 1$)

13. $m = 6$ and we must have $m \leq 2^r - r - 1$, so $r = 4$ and $n = 10$. One such matrix is:

$$H = \begin{bmatrix} 1 & 1 & 0 & 0 \\ 1 & 0 & 1 & 0 \\ 1 & 0 & 0 & 1 \\ 0 & 1 & 1 & 0 \\ 0 & 1 & 0 & 1 \\ 0 & 0 & 1 & 1 \\ 1 & 0 & 0 & 0 \\ 0 & 1 & 0 & 0 \\ 0 & 0 & 1 & 0 \\ 0 & 0 & 0 & 1 \end{bmatrix}$$

14. $n = 15$, $m = 11$, $r = 4$, and $m = 2^r - r - 1$, so the code is perfect. One possible H is

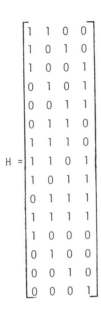

$$H = \begin{bmatrix} 1 & 1 & 0 & 0 \\ 1 & 0 & 1 & 0 \\ 1 & 0 & 0 & 1 \\ 0 & 1 & 0 & 1 \\ 0 & 0 & 1 & 1 \\ 0 & 1 & 1 & 0 \\ 1 & 1 & 1 & 0 \\ 1 & 1 & 0 & 1 \\ 1 & 0 & 1 & 1 \\ 0 & 1 & 1 & 1 \\ 1 & 1 & 1 & 1 \\ 1 & 0 & 0 & 0 \\ 0 & 1 & 0 & 0 \\ 0 & 0 & 1 & 0 \\ 0 & 0 & 0 & 1 \end{bmatrix}$$

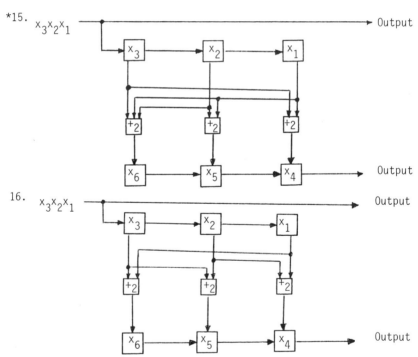

*15. $x_3 x_2 x_1$

16. $x_3 x_2 x_1$

17. $x_4x_3x_2x_1$

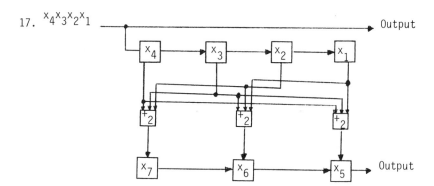

$0010 \longrightarrow 0010111$
$0101 \longrightarrow 0101110$

18. H' has no row of all 0's and no two rows alike. Also, three rows cannot sum to 0_{r+1} because the r+1$\underline{st}$ element is the sum of three 1's, hence equals 1. Therefore the minimum distance is ≥ 4. The rows which correspond to binary representations of 1,2,4, and 7 in H add to 0_{r+1} in H', so the minimum distance is 4.

19. For example, $0100 \longrightarrow 0100011$ by the calculation

$$
\begin{array}{r}
x^2 + x + 1 \\
x^3 + x^2 + 1 \overline{\smash{\big)}\, x^5 } \\
\underline{x^5 + x^4 + x^2} \\
x^4 + x^2 \\
\underline{x^4 + x^3 + x} \\
x^3 + x^2 + x \\
\underline{x^3 + x^2 + 1} \\
x + 1
\end{array}
$$

and $0110 \longrightarrow 0110100$ by the calculation

$$
\begin{array}{r}
x^2 \\
x^3 + x^2 + 1 \overline{\smash{\big)}\, x^5 + x^4 } \\
\underline{x^5 + x^4 + x^2} \\
x^2
\end{array}
$$

The remaining answers agree with Practice 8.20, and are obtained by similar calculations.

139

Section 8.2
*1. 0011010
   1100101
   0100011 (code word)
2. 11010
   0̂1111
   11010 (at least 2 errors have occurred)
3. Suppose X and Y are in the same coset of $\mathscr{C}$ in Z_2^n. Then $X = Y + C$ where $C \in \mathscr{C}$, and $X \cdot H = (Y + C)H = Y \cdot H + C \cdot H$. Because C is a code word, $C \cdot H = 0_r$ and $X \cdot H = Y \cdot H$. Now suppose $X \cdot H = Y \cdot H$. Then $(X - Y)H = 0_r$ and $X - Y$ is a code word C. But if $X - Y = C$, then $X = Y + C$, and X and Y are in the same coset.
4. The rows of H are r-tuples which are binary representations of the integers $1,\ldots,2^{r-1},\ldots,n$, $n < 2^r$. There are 2^r cosets, and 2^r syndromes, which consist of the set of all possible r-tuples; this set is the binary representation of all digits from 0 to $2^r - 1$. The coset leader for the syndrome 0_r in 0_n. Any syndrome which represents an integer between 1 and 2^{r-1} can be obtained by capturing the single row of H which represents the same integer, so a coset leader of weight 1 exists. For a syndrome which represents an integer between $2^{r-1} + 1$ and $2^r - 1$, the representation can be obtained by adding the representation of 2^{r-1} to the representation of an integer between 1 and $2^{r-1} - 1$. This requires adding together two rows of H, so the coset leader has weight 2.
*5. If only a single error has occurred, the coset leader has weight 1. If the 1 occurs in the $k^{\underline{th}}$ component, then the syndrome is the $k^{\underline{th}}$ row of H, which is the binary representation of k. Thus computing the syndrome of X gives the binary representation of the single component in X which should be changed.

6. a)

Coset leaders								Syndrome
000000	001101	010011	100111	110100	101010	011110	111001	000
000001	001100	010010	100110	110101	101011	011111	111000	001
000010	001111	010001	100101	110110	101000	011100	111011	010
010000	011101	000011	110111	100100	111010	001110	101001	011
000100	001001	010111	100011	110000	101110	011010	111101	100
001000	000101	011011	101111	111100	100010	010110	110001	101
000110	001011	010101	100001	110010	101100	011000	111111	110
100000	101101	110011	000111	010100	001010	111110	011001	111

b) 010011
110100
011110
110010; at least 2 errors have occurred, not decoded

*7. a)

Coset leaders	Syndromes
000000	000
000001	001
000010	010
001000	011
000100	100
100000	101
010000	110
100010	111

b) 101110
110011
110011
111111; at least 2 errors have occurred, not decoded

8. a) $$H^* = \begin{bmatrix} 0 & 0 & 1 \\ 0 & 1 & 0 \\ 0 & 1 & 1 \\ 1 & 0 & 0 \\ 1 & 0 & 1 \\ 1 & 1 & 0 \end{bmatrix}$$

b) To encode, we must have

$$[x_1 x_2 x_3 x_4 x_5 x_6] \begin{bmatrix} 0 & 0 & 1 \\ 0 & 1 & 0 \\ 0 & 1 & 1 \\ 1 & 0 & 0 \\ 1 & 0 & 1 \\ 1 & 1 & 0 \end{bmatrix} = [000]$$

This gives the system of equations

$$x_4 + x_5 + x_6 = 0$$
$$x_2 + x_3 + x_6 = 0$$
$$x_1 + x_3 + x_5 = 0$$

or

$$x_4 = -x_5 - x_6$$
$$x_2 = -x_3 - x_6$$
$$x_1 = -x_3 - x_5$$

We therefore encode a 3-tuple by taking its components to be x_3, x_5, and x_6, respectively, and then computing the values of x_4, x_2, and x_1. Thus
000 → 000000
001 → 010101
010 → 100110
100 → 111000
110 → 011110
101 → 101101
011 → 110011
111 → 001011

c) $X \cdot H = 011$, the binary representation for 3. Changing the third component of X, we decode to 010101.

9.

Coset leaders	Syndromes	Coset leaders	Syndromes
000000000 →	0000	000001000 →	1000
000000001 →	0001	000100000 →	1001
000000010 →	0010	000001010 →	1010
000000011 →	0011	000100010 →	1011
000000100 →	0100	000010000 →	1100
001000000 →	0101	100000000 →	1101
000000110 →	0110	000010010 →	1110
010000000 →	0111	010001000 →	1111

*10. a) One possible H is

$$H = \begin{bmatrix} 1 & 1 & 0 & 0 \\ 1 & 0 & 1 & 0 \\ 1 & 0 & 0 & 1 \\ 0 & 1 & 0 & 1 \\ 0 & 0 & 1 & 1 \\ 0 & 1 & 1 & 0 \\ 1 & 1 & 1 & 0 \\ 1 & 1 & 0 & 1 \\ 1 & 0 & 1 & 1 \\ 0 & 1 & 1 & 1 \\ 1 & 1 & 1 & 1 \\ 1 & 0 & 0 & 0 \\ 0 & 1 & 0 & 0 \\ 0 & 0 & 1 & 0 \\ 0 & 0 & 0 & 1 \end{bmatrix}$$

b) The syndrome for 011000010111001 is 1111. The coset leader is 000000000010000, so the word is decoded as 011000010101001.

11. a)

$$H = \begin{bmatrix} 0 & 0 & 1 & 1 \\ 0 & 1 & 0 & 1 \\ 0 & 1 & 1 & 0 \\ 0 & 1 & 1 & 1 \\ 1 & 0 & 0 & 0 \\ 0 & 1 & 0 & 0 \\ 0 & 0 & 1 & 0 \\ 0 & 0 & 0 & 1 \end{bmatrix}$$

b)

Coset leaders	Syndromes	Coset leaders	Syndromes
00000000	0000	00001000	1000
00000001	0001	00001001	1001
00000010	0010	00001010	1010
10000000	0011	10001000	1011
00000100	0100	00001100	1100
01000000	0101	01001000	1101
00100000	0110	00101000	1110
00010000	0111	00011000	1111

    c) 01100011
　　　 10010100

　　　 00111011; at least two errors have occurred, not decoded

12. Points to include: a table of coset leaders and syndromes must be available; since n is fairly small, this can be generated by "brute force" from the given matrix, as in Example 6.28. Coset leaders of weight 2 can be flagged. A received word is then located in the table and decoded by adding its coset leader, or flagged.

13.   010
　　　001
　　　111
　　　000

CHAPTER 9

Section 9.1

*1. a) 0001111110 b) aaacaaaa c) 00100110

2. a) 110100, 111010 b) none c) $0a_1a_2a_3a_4a_5$

3.

Present state	Next state Present input		Output
	0	1	
s_0	s_1	s_1	0
s_1	s_2	s_1	1
s_2	s_2	s_0	0

Output is 010010

4.

Present state	Next state Present input		Output
	0	1	
s_0	s_3	s_0	1
s_1	s_2	s_0	0
s_2	s_2	s_1	1
s_3	s_1	s_2	0

Output is 11101011

*5.

Present state	Next state Present input		Output
	0	1	
s_0	s_1	s_2	a
s_1	s_2	s_3	b
s_2	s_2	s_1	c
s_3	s_2	s_3	b

Output is abbcbb

6.

Present state	Next state Present input			Output
	a	b	c	
s_0	s_0	s_1	s_2	0
s_1	s_0	s_2	s_2	1
s_2	s_0	s_2	s_1	0

Output is 0000010

7.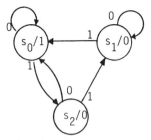

Output is 101110

8.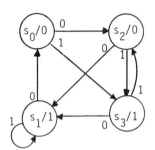

Output is 00111

*9.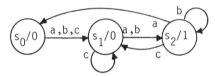

Output is 0001101

10.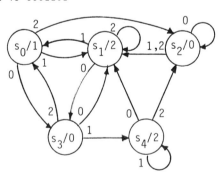

Output is 102012

11.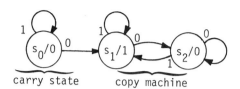

12. a)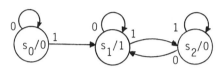

   b) 0100(0)
      0101(0)

*13. a)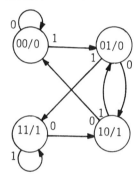

   b) The length of time required to remember a given input grows without bound and eventually would exceed the number of states.

*14.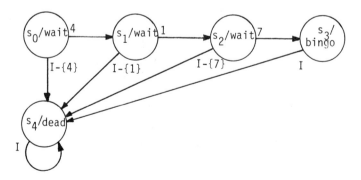

15.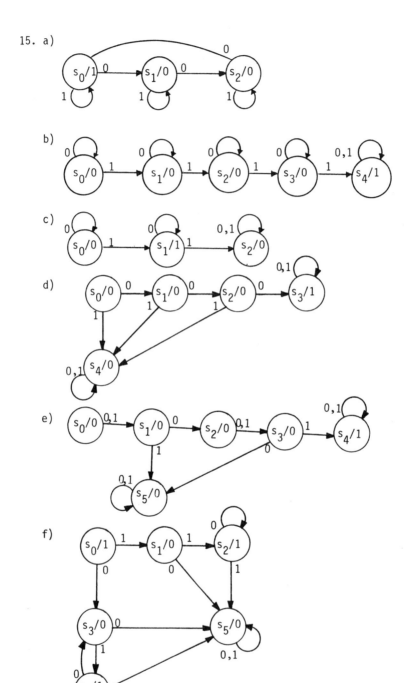

g)

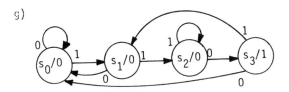

h)

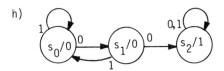

16. The object is to recognize the substring βcon.

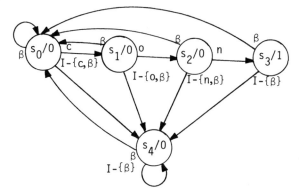

17. a)

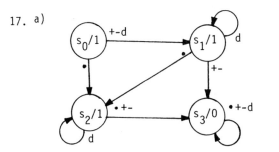

b)

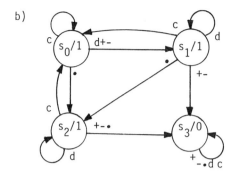

c) sd* or sdd*·dd* or d* or dd*·dd*
There should be extra states around · to guarantee at least one d.

*18. Once a state is revisited, behavior will be periodic since there is no choice of paths from a state. The maximum number of inputs which can occur before this happens is n - 1 (visiting all n states before repeating). The maximum length of a period is n (output from all n states, with the last state returning to s_0).

19. Points to include: the state table description will presumably be stored as an array structure; for each input symbol, the entry in the last column of the present row represents output, and the next row must be found by looking in the appropriate column of the present row.

20. B successfully receives message 0, now expects message 1, B's acknowledgment is on the channel

*21. A successfully receives acknowledgment of message 0 and then sends message 1

22. A has sent message 1 but it gets lost

23. B's acknowledgment of message 1 gets lost

24. A times out and retransmits message 1

25. B receives retransmitted message 1 and sends acknowledgment

26. A's retransmission of message 1 gets lost

Section 9.2

1. Corresponding states have the same output. There are eight cases to check:

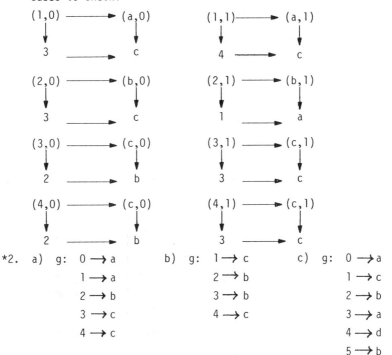

*2. a) g: 0 → a   b) g: 1 → c   c) g: 0 → a
       1 → a   2 → b   1 → c
       2 → b   3 → b   2 → b
       3 → c   4 → c   3 → a
       4 → c   4 → d
                                     5 → b

*3. a) States of M/g are [0] = {0,1}, [2] = {2}, [3] = {3,4}.

isomorphism:
[0] → a
[2] → b
[3] → c

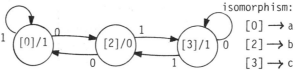

c) States of M/g are:
   [0] = {0,3}
   [1] = {1}
   [2] = {2,5}
   [4] = {4}

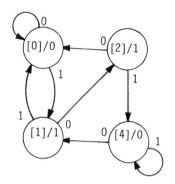

isomorphism: [0] → a
              [1] → c
              [2] → b
              [4] → d

4. a) A = {s_0, s_5}, B = {s_1, s_3}, C = {s_2, s_4}

Present state	Next state Present input		Output
	0	1	
A	B	A	1
B	C	B	1
C	A	A	0

g: s_0 → A
    s_1 → B
    s_2 → C
    s_3 → B
    s_4 → C
    s_5 → A

b) A = {s_0}, B = {s_2, s_3}, C = {s_1, s_4}

Present state	Next state Present input			Output
	a	b	c	
A	B	C	C	0
B	B	C	B	0
C	B	A	C	1

g: s_0 → A
    s_1 → C
    s_2 → B
    s_3 → B
    s_4 → C

5. A = {0,3} B = {1} C = {2}

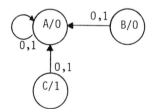

g: 0 → A
    1 → B
    2 → C
    3 → A

6. M simulates M by the identity function; symmetry obviously holds; if M and M' are equivalent, and M' and M" are equivalent, then M and M" are equivalent by using composition of the simulation functions. Each equivalence class consists of machines all of which simulate each other.

7. For example,

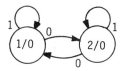 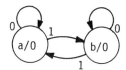

Section 9.3

1. a) 1*000* *b) 01* ∨ (110)* c) 11*(01)*
2. a) 10*1 b) (00)* ∨ 10* c) 1* ∨ (010)*
3. *a) 01*(001*)* b) 0*(0 ∨ 01)*
4. a) 0(0 ∨ 1)*1 b) 1*01*(01*0)*1* *c) 100*1
   d) (0 ∨ 1)*0(0 ∨ 1)* e) b*(abbb*)* f) 1*01*01*
*5. a) Yes b) No c) No d) Yes
6. a) L(L ∨ d)* where L stands for any letter, d stands for any digit
   b) dd*(+ ∨ -)dd*

7. In disconnecting a state (s_m or s_n), arcs from remaining states to that state are also removed; this may leave a remaining state with no next state for a particular input symbol.

8. a) (0 ∨ 1)1

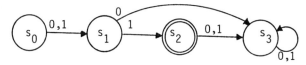

(This is deterministic)

b) (0 ∨ 1)*(00 ∨ 11)

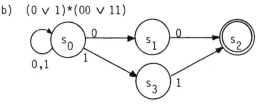

153

c) (0 ∨ 1)*111(0 ∨ 1)*

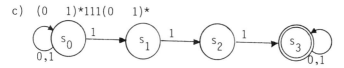

9.*a)

M' Present state	Next state Present input		Output
	0	1	
$\{s_0\}$	$\{s_0,s_1\}$	$\{s_1\}$	1
$\{s_1\}$	$\{s_0\}$	$\{s_0,s_1\}$	0
$\{s_0,s_1\}$	$\{s_0,s_1\}$	$\{s_0,s_1\}$	1

(0* ∨ (0 ∨ 1)1*(0 ∨ 1))*

b)

M' Present state	Next state Present input		Output
	0	1	
$\{s_0\}$	$\{s_1,s_2\}$	$\{s_0\}$	1
$\{s_1\}$	$\{s_0,s_1\}$	$\{s_2\}$	0
$\{s_2\}$	$\{s_2\}$	$\{s_2\}$	0
$\{s_0,s_1\}$	$\{s_0,s_1,s_2\}$	$\{s_0,s_2\}$	1
$\{s_1,s_2\}$	$\{s_0,s_1,s_2\}$	$\{s_2\}$	0
$\{s_0,s_2\}$	$\{s_1,s_2\}$	$\{s_0,s_2\}$	1
$\{s_0,s_1,s_2\}$	$\{s_0,s_1,s_2\}$	$\{s_0,s_2\}$	1

(1 ∨ 00*0)* or (1*(000*)*)*

c)

M' Present state	Next state Present input		Output
	0	1	
$\{s_0\}$	$\{s_2\}$	$\{s_1,s_3\}$	0
$\{s_1\}$	$\{s_2\}$	$\{s_1\}$	1
$\{s_2\}$	$\{s_2\}$	$\{s_2\}$	0
$\{s_3\}$	$\{s_4\}$	$\{s_2\}$	0
$\{s_4\}$	$\{s_2\}$	$\{s_5\}$	0
$\{s_5\}$	$\{s_2\}$	$\{s_3\}$	1
$\{s_1,s_3\}$	$\{s_2,s_4\}$	$\{s_1,s_2\}$	1
$\{s_1,s_2\}$	$\{s_2\}$	$\{s_1,s_2\}$	1
$\{s_2,s_4\}$	$\{s_2\}$	$\{s_2,s_5\}$	0
$\{s_2,s_5\}$	$\{s_2\}$	$\{s_2,s_3\}$	1
$\{s_2,s_3\}$	$\{s_2,s_4\}$	$\{s_2\}$	0

(11*) ∨ (101)*

10. a)

Present state	Next state Present input 0	1	Output
$\{s_0\}$	$\{s_2\}$	$\{s_1\}$	0
$\{s_1\}$	$\{s_2,s_3\}$	$\{s_3\}$	1
$\{s_2\}$	$\{s_3\}$	$\{s_1,s_3\}$	0
$\{s_3\}$	$\{s_3\}$	$\{s_3\}$	0
$\{s_1,s_3\}$	$\{s_2,s_3\}$	$\{s_3\}$	1
$\{s_2,s_3\}$	$\{s_3\}$	$\{s_1,s_3\}$	0

$01(01)^* \lor 1(01)^*$

b)

Present state	Next state Present input 0	1	Output
$\{s_0\}$	$\{s_0\}$	$\{s_1,s_2\}$	0
$\{s_1\}$	$\{s_3\}$	$\{s_3\}$	1
$\{s_2\}$	$\{s_3\}$	$\{s_1\}$	0
$\{s_3\}$	$\{s_3\}$	$\{s_3\}$	0
$\{s_1,s_2\}$	$\{s_3\}$	$\{s_1,s_3\}$	1
$\{s_1,s_3\}$	$\{s_3\}$	$\{s_3\}$	1

$0^*(1 \lor 11)$

11. a) Proof is by induction on the length of the regular expression. For the base step, if $A = \phi, \lambda$, or i, then $A^R = \phi, \lambda$, or i. Assume that for all expressions of length $\leq k$, A regular $\Rightarrow A^R$ regular. Let A be a regular expression of length $k + 1$. If $A = BC$, where B and C are regular, then B^R and C^R are regular by inductive hypothesis, and $A^R = C^R B^R$ so A^R is regular. Similarly, if $A = B \lor C$, then $A^R = B^R \lor C^R$ (regular), and if $A = B^*$, then $A^R = (B^R)^*$ (regular).

b) No, cannot write a regular expression for this set.

12. a) $A^R = 0^*001^*$. The nondeterministic machine is

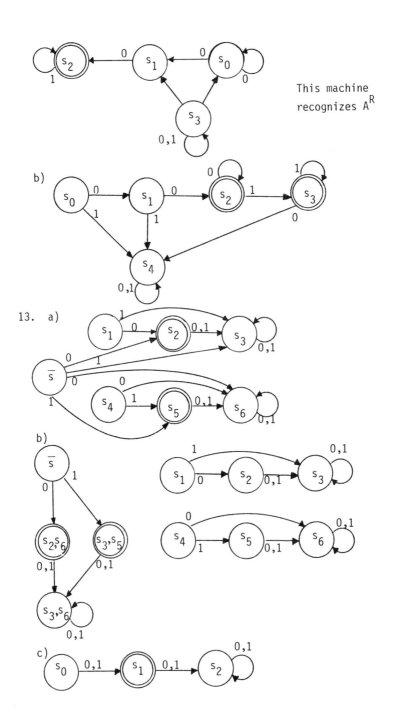

This machine recognizes A^R

*14. a)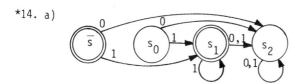

b)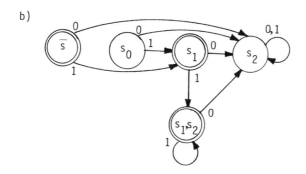

c)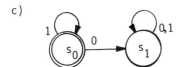

15. a)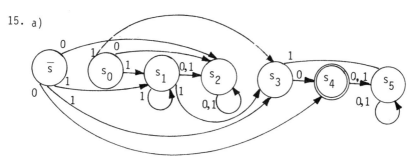

b) No answer

c)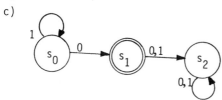

16. a)

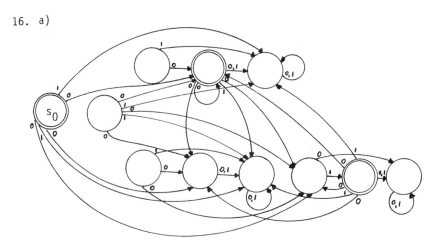

b) No answer

c)

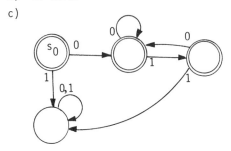

17. Let M be a finite-state machine recognizing A (M exists by Kleene's Theorem). Interchange final and nonfinal states of M to make a new machine M'. M' recognizes I* - A, so I* - A is regular by the other half of Kleene's Theorem.

CHAPTER 10

Section 10.1

*1. a) s_1 b) s_2

2. 5 is unreachable.

	1	2	3	4	5	6
d:	1	2	2	3	∞	1
s:	0	1	1	2	0	0

*3. A = {0}, B = {1,2,5}, C = {3,4}, D = {6}

Present state	Next state Present input 0	1	Output
A	C	D	1
B	C	B	0
C	B	A	1
D	C	B	1

4. A = {0,3,4}, B = {1}, C = {2,5}

Present state	Next state Present input 0	1	Output
A	C	A	1
B	C	C	0
C	B	A	0

5. A = {0}, B = {5}, C = {2}, D = {7,8}, E = {1,3}, F = {4,6}

Present state	Next state Present input 0	1	Output
A	E	C	0
B	F	D	0
C	E	F	0
D	D	E	0
E	C	E	1
F	B	F	1

159

6. A = {0,3}, B = {1,6}, C = {2,4}, D = {5,7}

Present state	Next state — Present input		Output
	0	1	
A	D	B	1
B	A	A	1
C	D	B	0
D	C	A	0

7. A = {0}, B = {2}, C = {1,4}, D = {3}, E = {5}

Present state	Next state — Present input		Output
	0	1	
A	C	D	0
B	E	C	0
C	B	C	1
D	C	B	2
E	C	A	2

*8. A = {0,2}, B = {1,3}, C = {4}

Present state	Next state — Present input			Output
	a	b	c	
A	B	C	A	1
B	C	A	B	0
C	B	A	A	0

9. A = {0,1,2,3}

Present state	Next state — Present input		Output
	0	1	
A	A	A	0

10. M is already minimized.

11. By the discussion in the text after Practice 10.5, the number of states of M* is ≤ the number of states of M**, and conversely. Therefore both machines have the same number of states. Also from the text, there is a mapping g: states of M* ⟶ states of

M** such that g is one-to-one. Therefore g is also onto. A state and its image have the same outputs for all inputs, including λ. If there is a state and its image which, under input symbol i, go to noncorresponding states, then these new states differ in output for some string α, so the original states differ in output for iα, a contradiction.

*12. a) True   b) False   c) True   d) False

Section 10.2

*1. Possible answer:

	d_1	d_2
s_0	0	0
s_1	0	1
s_2	1	0
s_3	1	1

x(t)	$d_1(t)$	$d_2(t)$	y(t)	$d_1(t+1)$	$d_2(t+1)$
0	0	0	0	1	0
1	0	0	0	1	1
0	0	1	1	0	0
1	0	1	1	0	1
0	1	0	0	0	1
1	1	0	0	1	1
0	1	1	1	0	1
1	1	1	1	1	0

$y(t) = d_1'd_2 + d_1 d_2 = d_2$
$d_1(t+1) = x'd_1'd_2' + xd_1'd_2' + xd_1'd_2 + xd_1 d_2 = d_1'd_2' + xd_1$
$d_2(t+1) = xd_1'd_2' + xd_1'd_2 + x'd_1 d_2' + xd_1 d_2' + x'd_1 d_2$
$\qquad = x(d_1' + d_2') + x'd_1$

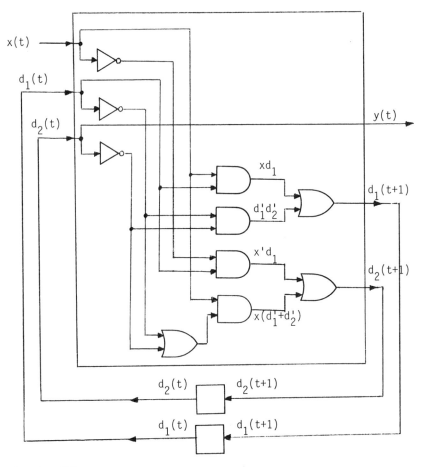

2. Possible answer:

	d_1	d_2	$x(t)$	$d_1(t)$	$d_2(t)$	$y(t)$	$d_1(t+1)$	$d_2(t+1)$
s_0	1	0	0	1	0	1	1	1
s_1	0	1	1	1	0	1	0	1
s_2	0	0	0	0	1	1	0	1
s_3	1	1	1	0	1	1	0	0
			0	0	0	0	0	0
			1	0	0	0	0	0
			0	1	1	0	1	0
			1	1	1	0	0	0

$$y(t) = d_1 d_2' + d_1' d_2$$
$$d_1(t+1) = x'd_1 d_2' + x'd_1 d_2 = x'd_1$$
$$d_2(t+1) = x'd_1 d_2' + x d_1 d_2' + x'd_1' d_2 = d_1 d_2' + x'd_1' d_2$$

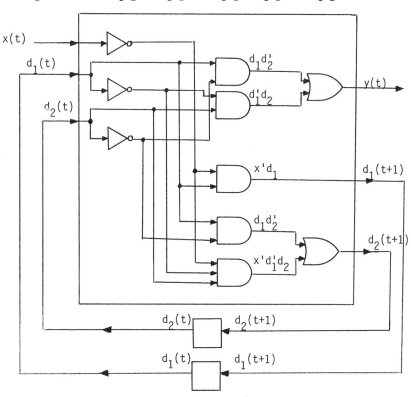

3. Possible answer:

Present state	Next state / Present input		Output
	0	1	
s_0	s_3	s_0	1
s_1	s_2	s_0	0
s_2	s_2	s_1	1
s_3	s_1	s_2	0

	d_1	d_2
s_0	0	0
s_1	0	1
s_2	1	0
s_3	1	1

$x(t)$	$d_1(t)$	$d_2(t)$	$y(t)$	$d_1(t+1)$	$d_2(t+1)$
0	0	0	1	1	1
1	0	0	1	0	0
0	0	1	0	1	0
1	0	1	0	0	0
0	1	0	1	1	0
1	1	0	1	0	1
0	1	1	0	0	1
1	1	1	0	1	0

$y(t) = d_1'd_2' + d_1d_2' = d_2'$

$d_1(t+1) = x'd_1'd_2' + x'd_1'd_2 + x'd_1d_2' + xd_1d_2 = x'(d_1' + d_2') + xd_1d_2$

$d_2(t+1) = x'd_1'd_2' + xd_1d_2' + x'd_1d_2$

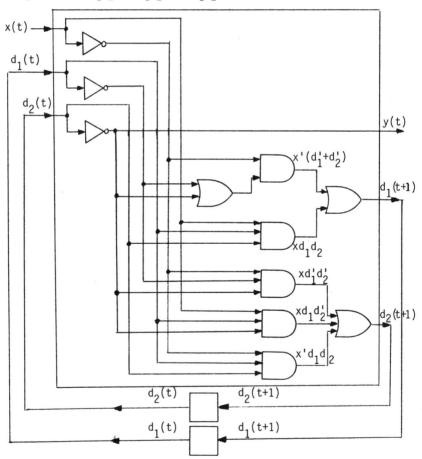

*4. Possible answer:

Present state	Next state Present input 0	1	Output
s_0	s_0	s_1	0
s_1	s_2	s_1	0
s_2	s_0	s_3	0
s_3	s_2	s_1	1

	d_1	d_2
s_0	0	0
s_1	0	1
s_2	1	0
s_3	1	1

$x(t)$	$d_1(t)$	$d_2(t)$	$y(t)$	$d_1(t+1)$	$d_2(t+1)$
0	0	0	0	0	0
1	0	0	0	0	1
0	0	1	0	1	0
1	0	1	0	0	1
0	1	0	0	0	0
1	1	0	0	1	1
0	1	1	1	1	0
1	1	1	1	0	1

$y(t) = d_1 d_2$

$d_1(t+1) = x'd_1'd_2 + xd_1d_2' + x'd_1d_2 = x'd_2 + xd_1d_2'$

$d_2(t+1) = xd_1'd_2' + xd_1'd_2 + xd_1d_2' + xd_1d_2 = x$

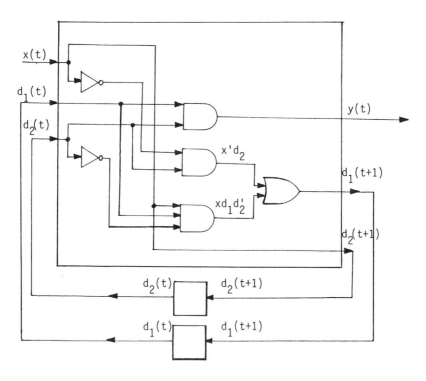

5. Possible answer:

output	y_1	y_2
0	0	0
1	0	1
2	1	0

	d_1	d_2
s_0	0	0
s_1	0	1
s_2	1	0
s_3	1	1

$x(t)$	$d_1(t)$	$d_2(t)$	$y_1(t)$	$y_2(t)$	$d_1(t+1)$	$d_2(t+1)$
0	0	0	0	0	0	0
1	0	0	0	0	1	1
0	0	1	0	1	0	0
1	0	1	0	1	1	0
0	1	0	0	1	1	1
1	1	0	0	1	1	1
0	1	1	1	0	0	1
1	1	1	1	0	1	1

$y_1(t) = d_1 d_2$

$y_2(t) = d_1'd_2 + d_1d_2'$
$d_1(t+1) = xd_1'd_2' + xd_1'd_2 + x'd_1d_2' + xd_1d_2' + xd_1d_2$
$\qquad = x + d_1d_2'$
$d_2(t+1) = xd_1'd_2' + x'd_1d_2' + xd_1d_2' + x'd_1d_2 + xd_1d_2$
$\qquad = d_1 + xd_2'$

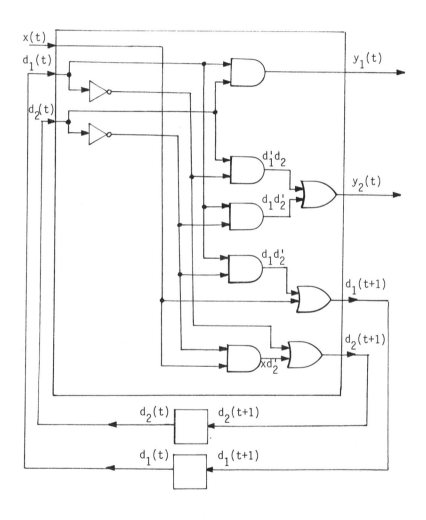

6. Possible answer:

| | Next state | |
| | Present input | |
Present state	00	01	10	11	Output
s_0	s_0	s_2	s_2	s_1	0
s_1	s_2	s_1	s_1	s_3	0
s_2	s_0	s_2	s_2	s_1	1
s_3	s_2	s_1	s_1	s_3	1

input	x_1	x_2
00	0	0
01	0	1
10	1	0
11	1	1

	d_1	d_2
s_0	0	0
s_1	0	1
s_2	1	0
s_3	1	1

$x_1(t)$	$x_2(t)$	$d_1(t)$	$d_2(t)$	$y(t)$	$d_1(t+1)$	$d_2(t+1)$
0	0	0	0	0	0	0
0	1	0	0	0	1	0
1	0	0	0	0	1	0
1	1	0	0	0	0	1
0	0	0	1	0	1	0
0	1	0	1	0	0	1
1	0	0	1	0	0	1
1	1	0	1	0	1	1
0	0	1	0	1	0	0
0	1	1	0	1	1	0
1	0	1	0	1	1	0
1	1	1	0	1	0	1
0	0	1	1	1	1	0
0	1	1	1	1	0	1
1	0	1	1	1	0	1
1	1	1	1	1	1	1

$y(t) = d_1$

$d_1(t+1)$ = (after Karnaugh map reduction) $x_1 x_2 d_2 + x_1' x_2' d_2$
$+ x_1 x_2' d_2' + x_1' x_2 d_2'$

$d_2(t+1)$ = (after Karnaugh map reduction) $x_1 d_2 + x_1 x_2 + d_2 x_2$

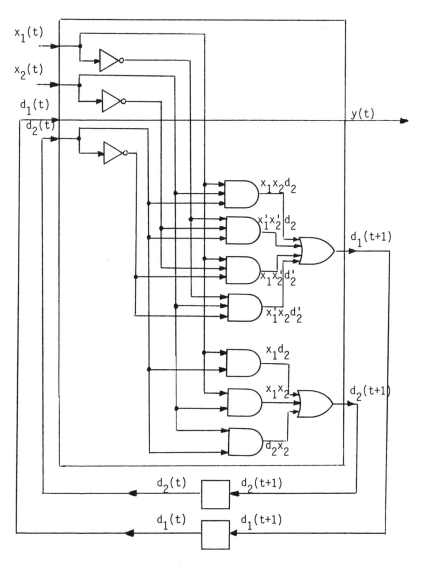

*7. Possible answer:

input	x_1	x_2
0	0	0
1	0	1
2	1	1

	d
s_0	0
s_1	1

$x_1(t)$	$x_2(t)$	$d(t)$	$y(t)$	$d(t+1)$
0	0	0	0	0
0	1	0	0	1
1	1	0	0	1
0	0	1	1	1
0	1	1	1	0
1	1	1	1	0
1	0	0	0	0
1	0	1	1	1

$y(t) = d$
$d(t+1) = x_2 d' + x_2' d$

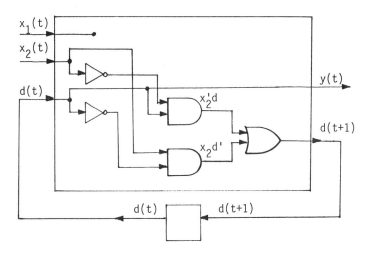

8. Possible answer:

	d_1	d_2	d_3
s_0	0	0	0
s_1	0	0	1
s_2	0	1	0
s_3	0	1	1
s_4	1	0	0

$x(t)$	$d_1(t)$	$d_2(t)$	$d_3(t)$	$y(t)$	$d_1(t+1)$	$d_2(t+1)$	$d_3(t+1)$
0	0	0	0	0	0	0	0
1	0	0	0	0	0	1	1
0	0	0	1	1	0	0	0
1	0	0	1	1	1	0	0
0	0	1	0	1	0	1	0
1	0	1	0	1	0	0	1
0	0	1	1	0	0	0	0
1	0	1	1	0	1	0	0
0	1	0	0	1	0	0	1
1	1	0	0	1	0	1	0
0	1	0	1	1	0	0	1
1	1	0	1	1	1	0	0
0	1	1	0	1	0	1	1
1	1	1	0	1	0	0	0
0	1	1	1	1	0	0	1
1	1	1	1	1	1	0	0

$y(t) = d_1 + d_2 d_3' + d_2' d_3$
$d_1(t + 1) = x d_3$
$d_2(t + 1) = d_3'(x d_2' + x' d_2)$
$d_3(t + 1) = x d_1' d_3' + x' d_1$

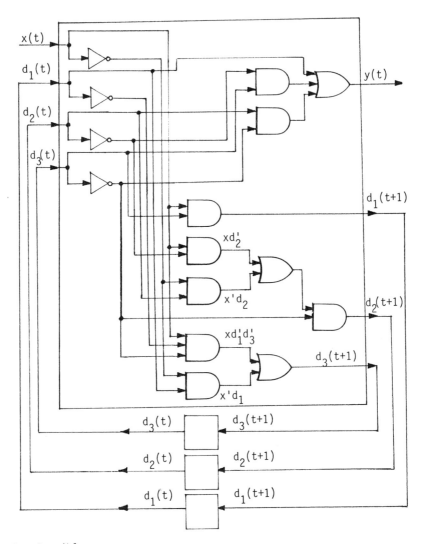

9. Possible answer:

	d_1	d_2
s_0	0	0
s_1	0	1
s_2	1	1

x(t)	$d_1(t)$	$d_2(t)$	y(t)	$d_1(t+1)$	$d_2(t+1)$
0	0	0	0	0	1
1	0	0	0	0	0
0	0	1	1	1	1
1	0	1	1	0	1
0	1	1	1	1	1
1	1	1	1	0	0
0	1	0	0	0	1
1	1	0	0	0	0

$y(t) = d_2$
$d_1(t+1) = x'd_2$
$d_2(t+1) = x' + d_1'd_2$

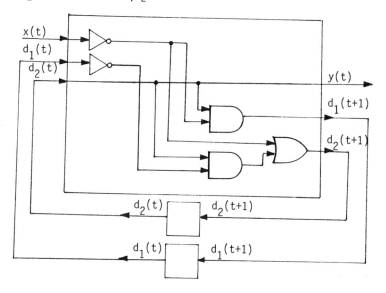

Section 10.3

*1. a) One example is

M_1	input 0 1	output
A = {0,1,3}	B A	0
B = {2,4}	A A	1

173

M_2	input 0 1	output
a = {0,2}	b c	0
b = {1,4}	a b	1
c = {3}	a a	1

g: (A,a) → 0   g': (0,0) → 0
 (A,b) → 1   (1,0) → 1
 (A,c) → 3   (0,1) → 1
 (B,a) → 2   (1,1) → 0
 (B,b) → 4

b) ((B,b),0) → (A,a) → 0
 ((A,a),1) → (A,c) → 3
 ((A,c),0) → (B,a) → 2

2. a) One example is:

M_1	input 0 1	output
A = {0,3,4}	B B	0
B = {1,2,5,6}	A B	1

M_2	input 0 1	output
a = {0,1}	b a	1
b = {2,3}	c d	1
c = {4,5}	c c	0
d = {6}	b c	0

g: (A,a) → 0   (B,a) → 1   g': (0,0) → 0
 (A,b) → 3   (B,b) → 2   (0,1) → 1
 (A,c) → 4   (B,c) → 5   (1,0) → 0
       (B,d) → 6   (1,1) → 0

b) ((A,c),0) → (B,c) → 5
 ((B,b),1) → (B,d) → 6
 ((B,d),0) → (A,b) → 3

3. a) One example is:

M_1	input 0	input 1	output
A = {0,3}	B	B	1
B = {1,2,4}	A	B	0

M_2	input 0	input 1	output
a = {0,2}	b	c	1
b = {3,4}	a	b	1
c = {1}	b	c	0

g: (A,a) → 0   g': (0,0) → 0
    (A,b) → 3   (0,1) → 0
    (B,a) → 2   (1,0) → 1
    (B,b) → 4   (1,1) → 1
    (B,c) → 1

b) ((A,a),0) → (B,b) → 4
   ((B,a),0) → (A,b) → 3
   ((A,b),1) → (B,b) → 4

4. a) One example is

M_1	input 0	input 1	output
A = {0,2,3}	B	A	0
B = {1,4,5}	B	B	1

M_2	input 0	input 1	output
a = {0,1}	a	b	0
b = {2,5}	b	c	1
c = {3,4}	c	a	x

g: (A,a) → 0   g': (0,0) → 0
    (A,b) → 2   (1,0) → 1
    (A,c) → 3   (0,1) → 1
    (B,a) → 1   (1,1) → 0
    (B,b) → 5   (0,x) → 0
    (B,c) → 4   (1,x) → 0

An extra symbol is required in the output alphabet for M_2.

b) $((A,c),0) \rightarrow (B,c) \rightarrow 4$
    $((B,a),1) \rightarrow (B,b) \rightarrow 5$
    $((A,a),1) \rightarrow (A,b) \rightarrow 2$

*5. a) One example is

M_1	input 0	input 1	output
A = {0,2,3}	B	A	A
B = {1,4,5}	A	B	B

M_2	(A,0)	(A,1)	(B,0)	(B,1)	output
a = {0,1}	a	b	a	b	a
b = {2,4}	c	c	c	a	b
c = {3,5}	b	b	c	a	c

g: (A,a) → 0   g': (A,a) → 0
    (A,b) → 2   (A,b) → 1
    (A,c) → 3   (A,c) → 0
    (B,a) → 1   (B,a) → 0
    (B,b) → 4   (B,b) → 1
    (B,c) → 5   (B,c) → 0

b) $((A,b),0) \rightarrow (B,c) \rightarrow 5$
    $((B,b),1) \rightarrow (B,a) \rightarrow 1$
    $((A,c),1) \rightarrow (A,b) \rightarrow 2$

6. a) One example is

M_1	input 0	input 1	output
A = {0,1}	B	B	A
B = {2,3,4}	A	B	B

M_2	(A,0)	(A,1)	(B,0)	(B,1)	output
a = {0,2}	c	a	a	c	a
b = {1,3}	a	b	b	a	b
c = {4}	–	–	a	b	c

176

g: $(A,a) \to 0$ g': $(A,a) \to 1$
   $(A,b) \to 1$ $(A,b) \to 1$
   $(B,a) \to 2$ $(B,a) \to 0$
   $(B,b) \to 3$ $(B,b) \to 1$
   $(B,c) \to 4$ $(B,c) \to 0$

b) $((A,a),0) \to (B,c) \to 4$
   $((B,a),0) \to (A,a) \to 0$
   $((B,c),1) \to (B,b) \to 3$

7. a) One example is

M_1	input 0	input 1	output
A = {0,3,4}	B	A	A
B = {1,5}	C	B	B
C = {2,6}	A	C	C

M_2	(A,0)	(A,1)	(B,0)	(B,1)	(C,0)	(C,1)	output
a = {0,1,2}	b	c	b	b	b	b	a
b = {3,5,6}	a	a	a	a	c	a	b
c = {4}	b	b	–	–	–	–	c

g: $(A,a) \to 0$ g': $(A,a) \to 1$
   $(A,b) \to 3$ $(A,b) \to 1$
   $(A,c) \to 4$ $(A,c) \to 0$
   $(B,a) \to 1$ $(B,a) \to 0$
   $(B,b) \to 5$ $(B,b) \to 1$
   $(C,a) \to 2$ $(C,a) \to 0$
   $(C,b) \to 6$ $(C,b) \to 0$

b) $((C,a),0) \to (A,b) \to 3$
   $((A,b),1) \to (A,a) \to 0$
   $((C,a),1) \to (C,b) \to 6$

Section 10.4

1. Note that $|G_{i-1}/G_i| = [G_{i-1}:G_i]$ and that $G/\{i\} = G$. The result is a finite extension of Exercise 28, Section 7.5.

2. $\{0\} \subset \{0,3,6,9,12\} \subset Z_{15}$ and $\{0\} \subset \{0,5,10\} \subset Z_{15}$ are two composition series. The factor sets are
$\{0,3,6,9,12\}/\{0\} \approx Z_5$, $Z_{15}/\{0,3,6,9,12\} \approx Z_3$ and
$\{0,5,10\}/\{0\} \approx Z_3$, $Z_{15}/\{0,5,10\} \approx Z_5$

*3. There are many possibilities, for example,
$\{0\} \subset \{0,30\} \subset \{0,6,12,18,24,30,36,42,48,54\} \subset$
$\{0,2,4,6,\ldots,58\} \subset Z_{60}$
$\{0\} \subset \{0,20,40\} \subset \{0,10,20,30,40,50\} \subset \{0,5,10,20,\ldots,55\} \subset Z_{60}$
$\{0\} \subset \{0,30\} \subset \{0,15,30,45\} \subset \{0,5,10,20,\ldots,55\} \subset Z_{60}$

In each case the members of the factor set are isomorphic to Z_2, Z_2, Z_3, and Z_5.

4. By the assumption, G_{k-1} is a subgroup of Z and by Theorem 5.58, $G_{k-1} = nZ$ for some positive integer n. But then $2nZ$ is a proper subgroup of G_{k-1}, it is normal because $[nZ,+]$ is commutative, and $\{0\} \subset 2nZ$. Hence $\{0\}$ is not maximal in G_{k-1}, a contradiction.

5. *a)

S	λ	0	1	2		i	01	11	10
							0	1	00
0	0	1	2	2
1	1	2	1	2
2	2	2	2	2

Table for S_M:

∘	i	0	1	2
i	i	0	1	2
0	0	2	0	2
1	1	2	1	2
2	2	2	2	2

Not a group.

b)

S	λ	00	000	01	010	011
	i		0	1	2	3	4
			111			
			100			
			001		101	
		0	1	10	11	110
0	0	2	1	1	0	2
1	1	1	0	2	1	1
2	2	0	1	1	0	2

∘	i	0	1	2	3	4
i	i	0	1	2	3	4
0	0	i	1	2	3	4
1	1	2	3	4	1	2
2	2	1	3	4	1	2
3	3	4	1	2	3	4
4	4	3	1	2	3	4

Not a group

c)

	i	0	1	2	3	4
	000	110	111			
	11	011	010	101	100	001
S	λ	0	1	00	01	10
0	0	1	2	2	1	0
1	1	2	1	0	0	2
2	2	0	0	1	2	1

∘	i	0	1	2	3	4
i	i	0	1	2	3	4
0	0	2	3	i	4	1
1	1	4	i	3	2	0
2	2	i	4	0	1	3
3	3	1	0	4	i	2
4	4	3	2	1	0	i

Group (S_3)

6. a) g: i → 0
       0 → 1
       1 → 2
       2 → 2

   b) g: i → 0
       0 → 2
       1 → 1
       2 → 1
       3 → 0
       4 → 2

   c) g: i → 0
       0 → 1
       1 → 2
       2 → 2
       3 → 1
       4 → 0

7. S_M is already a monoid, so we need only show the existence of inverses. Let $x \in S_M$. Then x represents a string of input symbols that produces a permutation on the states of S. If $x = i$, then x is its own inverse. If $x \neq i$, then $x^m = x^n$, $m > n$, for some n because there are at most k! distinct members of S_M. In this case, $x^{m-n} = i$ and $x^{m-n-1} = x^{-1}$.

8. a)

	i	1
	0	
S	λ	1
0	0	1
1	1	0

Table for S_M:

∘	i	1
i	i	1
1	1	i

Table for $[Z_2, +_2]$:

$+_2$	0	1
0	0	1
1	1	0

isomorphism: i → 0
             1 → 1

b)

	i	0	1	2
	(0,0)			
S	(λ,λ)	(0,1)	(1,0)	(1,1)
(0,0)	(0,0)	(0,1)	(1,0)	(1,1)
(0,1)	(0,1)	(0,0)	(1,1)	(1,0)
(1,0)	(1,0)	(1,1)	(0,0)	(0,1)
(1,1)	(1,1)	(1,0)	(0,1)	(0,0)

Table for semigroup:

o	i	0	1	2
i	i	0	1	2
0	0	i	2	1
1	1	2	i	0
2	2	1	0	i

Table for $[Z_2 \times Z_2, +]$

+	(0,0)	(0,1)	(1,0)	(1,1)
(0,0)	(0,0)	(0,1)	(1,0)	(1,1)
(0,1)	(0,1)	(0,0)	(1,1)	(1,0)
(1,0)	(1,0)	(1,1)	(0,0)	(0,1)
(1,1)	(1,1)	(1,0)	(0,1)	(0,0)

isomorphism: $i \to (0,0)$
$0 \to (0,1)$
$1 \to (1,0)$
$2 \to (1,1)$

c) Let S_{M_1} have m elements and S_{M_2} have n elements. This means there are m distinct transformations on S_1, the set of states of M_1, produced by input strings, and there are n distinct transformations on S_2, the set of states of M_2. There are m x n distinct transformations on $S_1 \times S_2$, the set of states of the parallel connection, and these are determined component-wise (perhaps concatenating λ to balance the lengths of the input strings in the ordered pairs). Composition in the semigroup of the parallel machine is done componentwise and therefore reflects the behavior of $S_{M_1} \times S_{M_2}$.

*9. The present state of the cascade machine is $([3],2)$ because $3 = 2 +_6 1$. The next state of the cascade machine is

$([3 +_6 4], 2 +_6 \beta(4,[3]))$
$= ([1], 2 +_6 (1 +_6 4) +_6 (-(3 +_6 4)'))$
$= ([1], 2 +_6 5 +_6 (-1))$
$= ([1], 2 +_6 5 +_6 5)$
$= ([1], 0)$

and $\gamma([1], 0) = 1$ because $1 = 0 +_6 1$

10. The elements of $Z_6/\{0,3\}$ are $[0] = \{0,3\}, [1] = \{1,4\}$ and $[2] = \{2,5\}$; let 0,1,2 be the fixed coset representatives.

*a) The present state of the cascade machine is $([2],0)$ because $2 = 0 +_6 2$. The next state of the cascade machine is

$([2 +_6 5], 0 +_6 \beta(5,[2]))$
$= ([1], 0 +_6 (2 +_6 5) + (-(2 +_6 5)'))$
$= ([1], 0 +_6 1 +_6 (-1))$
$= ([1], 1 +_6 5)$
$= ([1], 0)$

and $\gamma([1],0) = 1$ because $1 = 0 +_6 1$

b) The present state of the cascade machine is $([4],3)$ because $4 = 3 +_6 1$. The next state of the cascade machine is

$([4 +_6 1], 3 +_6 \beta(1,[4]))$
$= ([5], 3 +_6 (1 +_6 1) +_6 (-(4 +_6 1)'))$
$= ([5], 3 +_6 2 +_6 (-2))$
$= ([5], 3 +_6 2 +_6 4)$
$= ([5],3)$

and $\gamma([5],3) = 5$ because $5 = 3 +_6 2$.

11. The elements of $Z_{15}/\{0,5,10\}$ are $[0] = \{0,5,10\}$, $[1] = \{1,6,11\}$, $[2] = \{2,7,12\}$, $[3] = \{3,8,18\}$, $[4] = \{4,9,14\}$; let $0,1,2,3,4$ be the fixed coset representatives.

a) The present state of the cascade machine is $([6],5)$ because $6 = 5 +_{15} 1$. The next state of the cascade machine is

$([6 +_{15} 7], 5 +_{15} \beta(7,[6]))$
$= ([13], 5 +_{15} (1 +_{15} 7) +_{15} (-(6 +_{15} 7)'))$
$= ([13], 5 +_{15} 8 +_{15} (-3))$
$= ([13], 5 +_{15} 8 +_{15} 12)$
$= ([13],10)$

and $\gamma([13],10) = 13$ because $13 = 10 +_{15} 3$.

b) The present state of the cascade machine is $([8],5)$ because $8 = 5 +_{15} 3$. The next state of the cascade machine is

$([8 +_{15} 11], 5 +_{15} \beta(11,[8]))$
$= ([4], 5 +_{15} (3 +_{15} 11) +_{15} (-(8 +_{15} 11)'))$
$= ([4], 5 +_{15} 14 +_{15} (-4))$
$= ([4], 5 +_{15} 14 +_{15} 11)$
$= ([4],0)$

and $\gamma([4], 0) = 4$ because $4 = 0 +_{15} 4$.

CHAPTER 11

Section 11.1

*1. a) halts with final tape $\cdots$ | b | 0 | 0 | 0 | 0 | 0 | b | $\cdots$
    b) does not change the tape and moves forever to the left
2. a) halts with final tape $\cdots$ | b | 1 | 1 | 0 | 1 | 0 | b | $\cdots$
    b) moves forever back and forth adding a 1 at each end of the nonblank portion of the tape.
3. Points to include: if a match can be found for the (present state, present symbol) pair among the first two components of T's quintuples, then the symbol must be adjusted, the position on the tape adjusted, and the next state adjusted according to the last three components. If no match can be found, T halts.
4. (0,0,1,0,R)
    (0,1,0,0,R)
5. One answer: State 1 is a final state
    (0,0,0,0,R)   passes over 0's
    (0,1,1,1,R)   first 1, go to final state, halt and accept
*6. One answer:
    State 2 is a final state
    (0,b,b,2,R)   blank tape or no more 1's
    (0,1,1,1,R)   has read odd number of 1's
    (1,1,1,0,R)   has read even number of 1's
7. One answer: State 3 is a final state
    (0,0,0,0,R) ⎫
    (0,1,1,1,R) ⎬ pass over 0's to first 1
    (1,0,0,1,R) ⎫
    (1,1,1,2,R) ⎬ pass over 0's to second 1
    (2,b,b,3,R) } end of string, halt and accept
8. One answer: State 3 is a final state
    (0,(,(,0,R) ⎫
    (0,),X,1,L) ⎬ looks for leftmost), replaces it with X
    (0,X,X,0,R) ⎭
    (0,b,b,2,L) } no more)
    (1,(,X,0,R) ⎫
    (1,X,X,1,L) ⎬ looks for matching (

        (2,b,b,3,R) ⎫ no more (, halts and accepts
        (2,X,X,2,R) ⎭

*9. One answer: State 9 is a final state
    (0,b,b,9,R) } accepts blank tape
    (0,0,0,0,R) ⎫
                    finds first 1, marks with X
    (0,1,X,1,R) ⎭
    (1,1,1,1,R) ⎫
                    searches right for 2's
    (1,Y,Y,1,R) ⎭
    (1,2,Y,3,R) ⎫
                    pair of 2's, marks with Y's
    (3,2,Y,4,L) ⎭
    (4,Y,Y,4,L) ⎫
    (4,X,X,4,L) ⎪
                    searches left for 0's
    (4,1,1,4,L) ⎪
    (4,Z,Z,4,L) ⎭
    (4,0,Z,5,L) ⎫
                    pair of 0's, marks with Z's
    (5,0,Z,6,R) ⎭
    (6,Z,Z,6,R) ⎫
    (6,X,X,6,R) ⎬ passes right to next 1
    (6,1,X,1,R) ⎭
    (6,Y,Y,7,R) } no more 1's
    (7,Y,Y,7,R) ⎫
                    no more 2's
    (7,b,b,8,L) ⎭
    (8,Y,Y,8,L) ⎫
    (8,X,X,8,L) ⎪
                    no more 0's, halts and accepts
    (8,Z,Z,8,L) ⎪
    (8,b,b,9,L) ⎭

10. One answer: State 7 is a final state.
    (0,0,b,1,R) } 0 is leftmost symbol
    (1,0,0,1,R) ⎫
    (1,1,1,1.R) ⎪
                    finds right end
    (1,*,*,1,R) ⎪
    (1,b,b,2,L) ⎭
    (2,0,b,3,L) } match, erases right symbol
    (3,0,0,3,L) ⎫
    (3,1,1,3,L) ⎪
                    finds left end and begins again
    (3,*,*,3,L) ⎪
    (3,b,b,0,R) ⎭

$(0,1,b,4,R)$ } 1 is leftmost symbol
$(4,0,0,4,R)$
$(4,1,1,4,R)$
$(4,*,*,4,R)$ } finds right end
$(4,b,b,5,R)$
$(5,1,b,3,L)$ } match, erases right symbol
$(0,*,*,6,R)$ } word left of * is empty
$(6,b,b,7,R)$ } word right of * is empty, halts and accepts

11. One answer: State 8 is a final state
$(0,0,b,1,R)$ } 0 read on left of w_1
$(1,0,0,1,R)$
$(1,1,1,1,R)$ } moves right to *
$(1,*,*,2,R)$
$(2,X,X,2,R)$ } passes over X's
$(2,1,1,8,R)$
$(2,b,b,8,R)$ } nonzero on left of w_2, halts and accepts
$(2,0,X,3,L)$ } left symbols match
$(3,X,X,3,L)$
$(3,*,*,4,L)$ } moves left to *
$(4,1,1,4,L)$
$(4,0,0,4,L)$ } finds leftmost symbol
$(4,b,b,0,R)$
$(0,1,b,5,R)$ } 1 read on left of w_1
$(5,0,0,5,R)$
$(5,1,1,5,R)$ } moves right to *
$(5,*,*,6,R)$
$(6,X,X,6,R)$ } passes over X's
$(6,0,0,8,R)$
$(6,b,b,8,R)$ } non-one on left of w_2, halts and accepts
$(6,1,X,3,L)$ } left symbols match
$(0,*,*,7,R)$ } word left of * is empty
$(7,X,X,7,R)$
$(7,0,0,8,R)$ } word right of * nonempty, halts and accepts
$(7,1,1,8,R)$
$(0,b,b,0,R)$ } w_1 initially empty

12. A modification of the machine for Exercise 10 above will do the job.

13. One approach uses the following general plan: Put a marker at the right end of the original string X_1 and build a new string X_2 to the right of the marker. Working from the lower order end of X_1, for each new symbol in X_1, put a block of that symbol on the end of X_2 twice as long as the previous block of symbols added to X_2. When all symbols in X_1 have been processed, erase X_1 and the marker. Working from left to right in X_2, replace any 0's with 1's from the end of the string until there are no 0's left in X_2. (My implementation of this approach required 23 states and 85 quintuples; can you improve upon this solution?)

14. One answer:

(0,1,X,1,R) } X's leftmost 1 of original string
(1,1,1,1,R)
(1,b,*,2,R)
(1,*,*,2,R) } carries 1 to right end of new string and tacks it on
(2,1,1,2,R)
(2,b,1,3,L)

(3,1,1,3,L)
(3,*,*,4,L)
(4,1,1,4,L) } locates leftmost 1 of original string
(4,X,X,0,R)

(0,*,*,5,L)
(5,X,1,5,L) } original string all copied, changes its X's back to 1's and halts
(5,b,b,6,R)

*15. $f(n_1,n_2,n_3) = \begin{cases} n_2 + 1 & \text{if } n_2 > 0 \\ \text{undefined} & \text{if } n_2 = 0 \end{cases}$

16. (0,1,1,1,R)   this is an odd 1
    (1,1,1,0,R)   this is an even 1
    (0,b,1,2,R)   even number of 1's on tape, n odd, add 1 & halt
    (1,b,b,2,R)   odd number of 1's on tape, n even, halt

*17. (0,1,1,1,R)
    (1,b,1,4,R) } n = 0, add 1 and halt
    (1,1,1,2,R)
    (2,b,1,4,R) } n = 1, add additional 1 and halt

$\quad$ (2,1,1,3,R)
$\quad$ (3,1,b,3,R) $\}$ $n \geq 2$, erase extra 1's and halt
$\quad$ (3,b,b,4,R)

18. One answer:
$\quad$ (0,1,1,1,R)
$\quad$ (1,b,b,8,R) $\}$ $n = 0$, $2 \cdot 0 = 0$

$\quad$ (1,1,1,2,R)
$\quad$ (2,1,1,2,R) $\}$ $n > 0$, finds end of $\bar{n}$
$\quad$ (2,b,b,3,L)

$\quad$ (3,1,X,4,R)
$\quad$ (4,X,X,4,R)
$\quad$ (4,1,1,4,R) $\}$ changes 1 to X, adds 1 at right end of string
$\quad$ (4,b,1,5,L)

$\quad$ (5,1,1,5,L)
$\quad$ (5,X,X,6,L)
$\quad$ (6,X,X,6,L) $\}$ goes left to next 1 of $\bar{n}$
$\quad$ (6,1,X,4,R)

$\quad$ (6,b,b,7,R)
$\quad$ (7,X,1,7,R) $\}$ $\bar{n}$ is doubled, changes X's to 1's

$\quad$ (7,1,1,7,R)
$\quad$ (7,b,b,8,L) $\}$ finds right end, erases extra 1, halts
$\quad$ (8,1,b,9,L)

19. One answer:
$\quad$ (0,1,1,1,R) $\}$ ignores leading 1

$\quad$ (1,1,X,2,R)
$\quad$ (2,1,X,3,R) $\}$ counts XX1
$\quad$ (3,1,1,1,R)

$\quad$ (2,b,b,2,R)
$\quad$ (3,b,b,3,R) $\}$ $3 \nmid n$ so moves forever right

$\quad$ (1,b,b,4,L) $\}$ $3 \mid n$
$\quad$ (4,b,b,8,L) $\}$ $n = 0$

$(4,1,1,4,L)$
$(4,X,X,5,L)$
$(5,X,X,5,L)$
$(5,1,X,6,R)$ collects 1's at right end
$(6,X,X,6,R)$
$(6,1,1,6,R)$
$(6,b,1,4,L)$
$(5,b,b,7,R)$
$(7,X,b,7,R)$ 1's all collected, erases X's and halts
$(7,1,1,8,L)$

*20. One answer:
$(0,1,b,1,R)$ erases one extra 1
$(1,*,b,3,R)$ $n_1 = 0$
$(1,1,b,2,R)$
$(2,1,1,2,R)$ $n_1 > 0$, replaces * with leftmost 1 of $\overline{n_1}$, halts
$(2,*,1,3,R)$

21. One answer:
$(0,1,1,0,R)$
$(0,*,*,0,R)$
$(0,b,b,1,L)$ move to right end of 1's for n_2
$(0,X,X,1,L)$
$(1,1,X,2,L)$ X's rightmost 1 of n_2
$(2,1,1,2,L)$
$(2,*,*,2,L)$
$(2,b,b,3,R)$ move to left end of 1's for n_1, X leftmost 1
$(2,X,X,3,R)$
$(3,1,X,0,R)$
$(3,*,X,4,L)$ $n_1 < n_2$
$(4,X,X,4,L)$
$(4,b,1,5,R)$
$(5,X,b,5,R)$ write 0 on tape and halt
$(5,1,b,5,R)$
$(5,b,b,9,R)$
$(1,*,*,6,R)$ all of n_2 used, now write $n_1 - n_2$ on tape
$(6,X,X,6,R)$
$(6,b,b,7,L)$ erase n_2
$(7,X,b,7,L)$

$(7,*,1,8,L)$
$(8,1,1,8,L)$
$(8,X,b,8,L)$ } clean up $n_1 - n_2$ and halt
$(8,b,b,9,R)$

22. One answer:
    $(0,1,X,1,R)$ } changes first 1 of $\bar{n}_1$ to X
    $(1,1,1,1,R)$
    $(1,*,*,2,R)$
    $(2,X,X,2,R)$ } finds leftmost 1 of $\bar{n}_2$, changes to X
    $(2,1,X,3,L)$
    $(3,X,X,3,L)$
    $(3,*,*,4,L)$
    $(4,1,1,4,L)$ } finds leftmost 1 of $\bar{n}_1$, changes to X
    $(4,X,X,5,R)$
    $(5,1,X,1,R)$
    $(5,*,*,6,R)$ } $\bar{n}_1$ exhausted, $n_2 \geq n_1$
    $(6,X,1,6,R)$
    $(6,1,1,7,L)$ } changes X's in $\bar{n}_2$ to 1's
    $(6,b,b,7,L)$
    $(7,1,1,7,L)$
    $(7,*,b,7,L)$
    $(7,X,b,7,L)$ } erases $\bar{n}_1$ and halts
    $(7,b,b,10,L)$
    $(2,b,b,8,L)$ } $\bar{n}_2$ exhausted, $n_1 > n_2$
    $(8,X,b,8,L)$
    $(8,*,6,9,L)$ } erases $\bar{n}_2$
    $(9,X,1,9,L)$
    $(9,b,b,10,L)$ } changes X's in $\bar{n}_1$ to 1's, halts

23. invoke T_1, invoke T_2

24. T erases the leading 1 of $\bar{n}_1$ and moves right; if * is immediately encountered, $n_1 = 0$, so move right erasing $*n_2$, finally resulting in a single 1, then halt ($0 \cdot n_2 = 0$). If * is not immediately encountered, move right to $\bar{n}_2$ and count the leading 1's; if there is only one 1 in $\bar{n}_2$, then $n_2 = 0$, go left erasing $\overline{n_1 - 1}*$ and halt ($n_1 \cdot 0 = 0$). If $n_2 \neq 0$, go left to check if $n_1 = 1$. If so, erase $\overline{n_1 - 1}*$ and halt ($1 \cdot n_2 = n_2$). Otherwise,

sweep back and forth by Xing out $\overline{n_1 - 1}$ and moving right to the to the left end of the rightmost block, then COPYING. This creates $\overline{n_1 - 1}$ copies of $\overline{n_2}$. Erase $\overline{n_1 - 1}$ and move right to the left end of the second block from the right. Do a succession of ADDS, moving the resulting string left to close up the gap in the tape. When only a single block of 1's remains on the tape, halt; the result is $\overline{n_1 \cdot n_2}$.

*25. a) T may run forever processing a given input string α, and we would be unable to test other strings.

b) Let T be a Turing machine which computes the function. Using copies T_1, T_2, ... of T, feed input into them, and check to see which have halted with a representation $\overline{m}$ on the tape; any m so represented is in the range set.

26. An algorithm is a finite string of symbols where the symbols come from a finite set, say the 26 letters of the English alphabet together with a finite number of punctuation marks and special symbols. The set of all such strings is equivalent to N by listing in some lexicographical ordering all strings of length 1, then all strings of length 2, etc. The algorithms are an infinite subset of this list, so have cardinality $\aleph_0$.

Section 11.2

1. For any Turing machine T, we can effectively create a machine T that acts like T but replaces any occurrences of m as a state symbol with a new state symbol; in addition, whenever T reaches a halting configuration, T* enters state m. Then T* enters state m when started on a tape containing α if any only if T halts on α. Assume that an algorithm P exists to solve the state problem. An algorithm to solve the halting problem is: given a (T,α) pair, create T*, and then apply P.

2. For any Turing machine T and string α, we can effectively replace any instances of the symbol s in α with a new symbol. We can also effectively create a machine T* that acts like T but replaces any instances of s with the same new symbol and in addition, whenever T reaches a halting configuration, T* prints s on the tape. Then T* prints s when started on a tape containing α if any only if T halts on α. Assume

that an algorithm P exists to solve the printing problem. An algorithm to solve the halting problem is: given a (T,α) pair, modify α and create $T*$, and then apply P.

3. For any Turing machine T we can effectively create a machine $T*$ that first erases its tape and then turns computation over to T. Then $T*$ halts on every tape if and only if T halts on a blank tape. Assume that there exists an algorithm P to solve the uniform halting problem. An algorithm to solve the blank tape halting problem is: given T, create $T*$ and apply P.

4. For any Turing machine T we can effectively create a machine $T*$ that acts as follows on any $\bar{n}$: $T*$ moves beyond $\bar{n}$ to blank tape and turns computation over to T (always moving $\bar{n}$ out of the way, if necessary); from any halting configuration of T, $T*$ erases the tape except for $\bar{n}$, moves back to the left of $\bar{n}$, and turns computation over to M. Then $T*(\bar{n}) = M(\bar{n})$ for all n if and only if T halts on a blank tape. Assume that an algorithm P exists to solve the equivalence problem. An algorithm to solve the blank tape halting problem is: given T, create $T*$ and apply P. (Note that we have assumed here that $M(\bar{n})$ is not everywhere undefined, because if it is, then $T*(\bar{n}) = M(\bar{n})$ for all n does not imply that T halts on a blank tape. But, can we effectively decide, given an arbitrary Turing machine M, whether its number-theoretic function of one variable is everywhere undefined?)

<u>Section 11.3</u>

*1. It requires $\bar{n} + 1$ moves to locate the blank beyond the $\bar{n}$ symbols, and to move left to the end of $\bar{n}$. The machine can then X out successive symbols of $\bar{n}$ and add 1's on the right end. Working from the "middle" outward, this requires, successively, $1 + 2 + 3 + \cdots + 2\bar{n}$ steps. To blank out n X's and change the remaining X to 1 requires $\bar{n} + 3$ more steps. The total number of steps is <u>thus</u> $\bar{n} + 1 + (1 + 2 + \cdots + 2\bar{n}) + \bar{n} + 3$
$= 2\bar{n} + 4 + \frac{2\bar{n}(2\bar{n} + 1)}{2} = 2\bar{n}^2 + 3\bar{n} + 4 = 2(n + 1)^2 + 3(n + 1) + 4$,
which is of order n^2.

2.-6. In each case all the possible candidates for success can be processed in parallel, that is, an NDTM can "guess" truth assignments (or subgraphs of G or sets from C or node colorings or bin assignments) and then test to see if any of the guesses produce a solution.

CHAPTER 12

Section 12.1

*1. a) $L(G) = \{a\}$
   b) $L(G) = \{010101, 010111, 011101, 011111, 110101, 110111, 111101, 111111\}$
   c) $L(G) = 0(10)*$
   d) $L(G) = 0*1111*$

2. a) (c) is regular, (b) and (d) are context-free
   b) For example:
      for (a): $G = (V, V_T, S, P)$ where $V = \{a, S\}$, $V_T = \{a\}$, and $P = \{S \rightarrow a\}$.
      for (b): $G = (V, V_T, S, P)$ where $V = \{0, 1, A, B, C, D, E, F, G, S\}$, $V_T = \{0, 1\}$, and P consists of the productions

      $S \rightarrow 0A$ $C \rightarrow 1D$
      $S \rightarrow 1A$ $D \rightarrow 1E$
      $A \rightarrow 1B$ $D \rightarrow 1F$
      $A \rightarrow 1C$ $E \rightarrow 0G$
      $B \rightarrow 0D$ $F \rightarrow 1G$
                             $G \rightarrow 1$

      for (d): $G = (V, V_T, S, P)$ where $V = \{0, 1, A, B, S\}$, $V_T = \{0, 1\}$, and P consists of the productions

      $S \rightarrow 0S$ $B \rightarrow 1B$
      $S \rightarrow 1A$ $B \rightarrow 1$
      $A \rightarrow 1B$

*3. $L(G) = aa*bb*$. G is context-sensitive. An example of a regular grammar that generates $L(G)$ is $G' = (V, V_T, P, S)$ where $V = \{a, b, A, B, S\}$, $V_T = \{a, b\}$, and P consists of the productions

   $S \rightarrow aA$ $A \rightarrow aA$ $B \rightarrow bB$
   $S \rightarrow aB$ $A \rightarrow aB$ $B \rightarrow b$

4. a) $<S> ::= \lambda | 0<A>$
      $<A> ::= 0<B>$
      $<B> ::= 0 | 0<C>$
      $<C> ::= 0<B>$
   b) $<S> ::= 0<A> | 1<A>$
      $<A> ::= 1<B><B>$
      $<B> ::= 01 | 11$

c) <S> ::= 0|0<A>
   <A> ::= 1<B>
   <B> ::= 0<A>|0
d) <S> ::= 0<S>|11<A>
   <A> ::= 1<A>|1

5. For example, $G = (V, V_T, S, P)$ where $V = \{(,),S\}$, $V_T = \{(,)\}$, and P consists of the productions
$$S \to \lambda$$
$$S \to (S)S$$

6. a) For example, $G = (V, V_T, S, P)$ where $V = \{a,b,A,B,S\}$, $V_T = \{a,b\}$, and P consists of the productions
$$S \to \lambda \qquad A \to a$$
$$S \to ASA \qquad B \to b$$
$$S \to BSB$$
$$S \to a$$
$$S \to b$$

b) Let $w \in L$. Then $w = w^R$, or $w^R = (w^R)^R$. Therefore all the members of L^R are palindromes.

c) Let $x \in w^*$. Then $x = \lambda$ (a palindrome), or x is the concatenation of k copies of w, $x = w^1 \cdots w^k$, $k \geq 1$. If $s = w^1 \cdots w^k$, then $x^R = (w^1 \cdots w^k)^R = (w^k)^R \cdots (w^1)^R = w \cdots w = x$. Therefore x is a palindrome.

*7. For example, $F = (V, V_T, S, P)$ where $V = \{0,1,A,S\}$, $V_T = \{0,1\}$ and P consists of the productions
$$S \to 01 \qquad A \to A0 \qquad A \to 0$$
$$S \to A01 \qquad A \to A1 \qquad A \to 1$$

8. For example, $G = (V, V_T, S, P)$ where $V = \{0,1,S,S_1\}$, $V_T = \{0,1\}$, and P consists of the production
$$S \to \lambda$$
$$S \to 01$$
$$S \to 0S_1 1$$
$$S_1 \to 0S_1 1$$
$$S_1 \to 01$$

9. For example, $G = (V, V_T, S, P)$ where $V = \{0, S, A, B, X\}$, $V_T = \{0\}$, and P consists of the productions

$S \rightarrow A0B$ $XB \rightarrow B$
$A0 \rightarrow A00X$ $A \rightarrow \lambda$
$X0 \rightarrow 00X$ $B \rightarrow \lambda$

10. For example, $G = (V, V_T, S, P)$ where $V = \{0, 1, A, B, S\}$, $V_T = \{0, 1\}$, and P consists of the productions

$S \rightarrow 0A$ $A \rightarrow 0AA$
$S \rightarrow 1B$ $B \rightarrow 0$
$A \rightarrow 1$ $B \rightarrow 0S$
$A \rightarrow 1S$ $B \rightarrow 1BB$

11. For example, $G = (V, V_T, S, P)$ where $V = \{0, 1, S\}$, $V_T = \{0, 1\}$, and P consists of the productions

$S \rightarrow 001$ $S \rightarrow 1S00$
$S \rightarrow 100$ $S \rightarrow 10S0$
$S \rightarrow 010$ $S \rightarrow 0S10$
$S \rightarrow 0S01$ $S \rightarrow 01S0$
$S \rightarrow 00S1$ $S \rightarrow SS$

*12. For example, $G = (V, V_T, S, P)$ where $V = \{0, 1, S, S_1\}$, $V_T = \{0, 1\}$, and P consists of the productions

$S \rightarrow \lambda$ $S_1 \rightarrow 1S_1 1$
$S \rightarrow S_1$ $S_1 \rightarrow 00$
$S_1 \rightarrow 0S_1 0$ $S_1 \rightarrow 11$

13. For example, $G = (V, V_T, S, P)$ where $V = \{0, 1, S, T, A, B, Y, Z, M, N, E\}$, $V_T = \{0, 1\}$, and P consists of the production

$S \rightarrow \lambda$
$S \rightarrow ATYE$
$S \rightarrow BTZE$
$T \rightarrow ATY$
$T \rightarrow BTZ$

$YE \rightarrow ME$ $ZE \rightarrow NE$
$YM \rightarrow MY$ $YN \rightarrow NY$
$ZM \rightarrow MZ$ $ZN \rightarrow NZ$
$TM \rightarrow TA$ $TN \rightarrow TB$
$AM \rightarrow AA$ $AN \rightarrow AB$
$BM \rightarrow BA$ $BN \rightarrow BB$

$$A \rightarrow 0$$
$$B \rightarrow 1$$
$$E \rightarrow \lambda$$
$$T \rightarrow \lambda$$

(The basic plan here is to generate wTw^R and then reverse w^R.)

14. For example, $G = (V, V_T, S, P)$ where $V = \{a,b,S,G,E,D,F,L,B,M,N,Q,R,Z,X,P,U\}$, $V_T = \{a,b\}$, and P consists of the productions

    $S \rightarrow GFDE$
    $D \rightarrow a$ ⎱ D generates any number of a's
    $D \rightarrow Da$ ⎰
    $Fa \rightarrow aLB$ ⎫
    $Ba \rightarrow aB$
    $Bb \rightarrow bB$
    $BE \rightarrow bMNE$
    $BN \rightarrow B$ ⎬ construct the same number of b's as a's
    $bM \rightarrow Mb$
    $aM \rightarrow Ma$
    $LM \rightarrow F$
    $Fb \rightarrow Qb$ ⎭
    $aQ \rightarrow RZ$ erase one a
    $Zb \rightarrow bLX$ ⎫
    $Xb \rightarrow bX$
    $XN \rightarrow PNa$ ⎬ copy same number of a's as b's on other side of b's
    $bP \rightarrow Pb$
    $LP \rightarrow Z$ ⎭
    $ZN \rightarrow QN$ ⎱ marks finish of a generation of a's
    $bQ \rightarrow Qb$ ⎰
    $RQ \rightarrow Q$ back at original a's, loop
    $GQ \rightarrow U$ all original a's have been erased
    $Ub \rightarrow U$ erase all b's
    $UN \rightarrow U$ erase end of B's
    $Ua \rightarrow aU$ leave a's
    $UE \rightarrow \lambda$ erase end marker

15. *a) 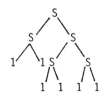   b)

c)   d)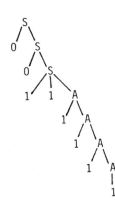

*16.

17. The set of productions for G' is formed from the set of productions for G as follows: For A and B nonterminals, whenever $A \overset{*}{\Rightarrow} B$ in G and $B \longrightarrow \alpha$ is a production in G with $|\alpha| \geq 2$, add the production $A \longrightarrow \alpha$ to the set, then eliminate all productions of the form $A \longrightarrow B$. For any productions of the form $A \longrightarrow a$, $a \in V_T$, add to the set of productions those obtained by replacing any A on the right of an existing production by a, then eliminate all productions of the form $A \longrightarrow a$. Eliminate $S \longrightarrow \lambda$. The remaining productions all have right side with length ≥ 2, and $L(G') \subseteq L(G)$. Only λ and a finite number of one-length words may have been eliminated, so $L(G) - L(G')$ is a finite set.

Section 12.2

*1. $G = (V, V_T, S, P)$ where $V = \{S, A, B, C, 0, 1\}$

V_T = {0,1}, and P consists of the productions

S → 1A	A → 0	B → 0	C → 0C
S → 0C	A → 0B	B → 0B	C → 1C
	A → 1C	B → 1C	

L(G) = 100*

2. G = (V,V_T,S,P) where V = {S,A,B,C,0,1}

V_T = {0,1}, and P consists of the productions

S → λ	B → 0B
S → 0	B → 1B
S → 0C	C → 0C
S → 1A	C → 1A
A → 1B	C → 0
A → 0C	
A → 0	

L(G) = (0 ∨ 10)*

3. G = (V,V_T,S,P) where V = {S,A,B,C,D,a,b,c}, V_T = {a,b,c}, and P consists of the productions

S → aA	B → c	C → aD
S → bB	B → aD	C → bD
S → cD	B → bD	C → cD
A → aS	B → cC	
A → bD	D → aD	
A → cD	D → bD	
	D → cD	

L(G) = (aa)*bc

4.

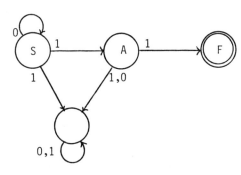

L(G) = 0*11

*5.

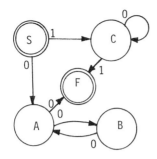

L(G) = λ ∨ 0(00)*0 ∨ 10*1

6.

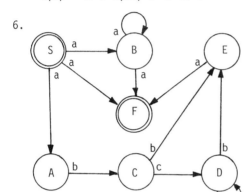

L(G) = abc*ba ∨ a*

*7. There are n words of length 1
     n·n words of length 2
     .
     .
     .
     n^k words of length k

So the total number of words is $n + n^2 + \cdots + n^k$

8. If L is not context-sensitive, erasing productions can exist, and we cannot state that W_i is the set of all words in L of length $\leq k$ which are derivable from S in no more than i steps. For example, if k = 4, a word w of length 4 may be derivable from S in no more than 6 steps but its predecessor in the derivation could have length >4, so it would not be in W_5, hence w would not be in W_6.

9. $W_0 = \{S\}$
    $W_1 = \{S,0,1A\}$
    $W_2 = \{S,0,1A,1B\}$
    $W_3 = \{S,0,1A,1B,11,10B1\}$
    $W_4 = \{S,0,1A,1B,11,10B1,1011\}$
    $W_5 = W_4$ so $1010 \notin L(G)$.

10. *a) $W_0 = \{S\}$
    $W_1 = \{S,a,b,bA\}$
    $W_2 = \{S,a,b,bA,bc\}$
    $W_3 = W_2$ so $cb \notin L(G)$.

    b) $W_0 = \{S\}$
    $W_1 = \{S,S+S,a,b,bA\}$
    $W_2 = \{S,S+S,a,b,bA,SSa,a+S,S+a,b+S,S+b,bc\}$
    $W_3 = W_2 \cup \{aSa,Saa,bSa,Sba,a+a,a+b,b+a,b+b\}$
    $W_4 = W_3 \cup \{aaa,aba,baa,bba\}$
    $W_5 = W_4$ so $bbc \notin L(G)$

11. S is a subset of $\{0,1\}^*$. Furthermore S is recursive: given an $x_i \in \{0,1\}^*$, we can effectively find i and G_i. Because G_i is context-sensitive, $L(G_i)$ is recursive and we can decide whether $x_i \in L(G_i)$ and thus whether $x_i \in S$. If S were context-sensitive, then $S = L(G_k)$ for some context-sensitive grammar G_k. Then $x_k \in S \Longleftrightarrow x_k \notin S$, a contradiction. Thus S is not context-sensitive.

12. a) Suppose that L is context-free, and let k be the constant of the pumping lemma. Let n be such that $|a^n b^n c^n| \geq k$. Then $a^n b^n c^n = w_1 w_2 w_3 w_4 w_5$. Because $|w_2 w_4| \geq 1$, not both w_2 and w_4 are empty. Neither w_2 nor w_4 can contain more than one symbol; if, for example, $w_2 = aab$ then $w_1 w_2^2 w_3 w_4^2 w_5$, which is in L, would contain the string aabaab, a contradiction. If w_2 contains only a's, then the word $w_1 w_2^2 w_3 w_4^2 w_5$ contains more than n a's; if w_4 is empty, contains only a's or contains only b's, then $w_1 w_2^2 w_3 w_4^2 w_5$ does not contain more than n c's, so does not belong to L, a contradiction, while if w_4 contains only c's, then $w_1 w_2^2 w_3 w_4^2 w_5$ does not contain more than n b's, a contradiction. The other cases are similar.

b) Suppose that L is context-free, and let k be the constant of the pumping lemma. Let n be such that $|a^{n^2}| \geq k$. Then $a^{n^2} = w_1 w_2 w_3 w_4 w_5$, or $w_1 = a^u$, $w_2 = a^v$, $w_3 = a^w$, $w_4 = a^x$, $w_5 = a^y$ with v and x not both zero. Then $w_1 w_2 w_3 w_4 w_5 = a^u a^v a^w a^x a^y = a^{u+v+w+x+y} \in L$, so $u + v + w + x + y$ is a perfect square, call it s. Using i = 2 in the pumping lemma, $u + 2v + w + 2x + y$ is a perfect square; using i = 3, $u + 3v + w + 3x + y$ is a perfect square, etc. Thus there exists a perfect square s and a fixed number $d = v + x > 0$ such that $s + jd$ is a perfect square for any $j \geq 0$. This is a contradiction, since perfect squares grow progressively farther apart.